THE VOICE OF THE MOUNTAINS

Radio and Anthropology

Alan O'Connor

University Press of America,® Inc.
Lanham · Boulder · New York · Toronto · Oxford

Copyright © 2006 by
University Press of America,® Inc.
4501 Forbes Boulevard
Suite 200
Lanham, Maryland 20706
UPA Acquisitions Department (301) 459-3366

PO Box 317
Oxford
OX2 9RU, UK

An earlier version of Chapter 3 was published in *Studies in Latin American Popular Culture*.
Appendices A and B are translated from *Materiales Para la Comunicación Popular* 9 (1987), published in Lima, Peru, by the Institute for Latin America (IPAL).

Library of Congress Control Number: 2006927250
ISBN-13: 978-0-7618-3537-0 (paperback : alk. paper)
ISBN-10: 0-7618-3537-7 (paperback : alk. paper)

Contents

Preface

I have wanted for some time to write something about anthropology and radio broadcasting. Rather than being welcomed as a dialogue between ethnography and communication studies, my efforts are in danger of being declared offside. The whistle blows and what can I say? This book is intended as a critique of the field of development communication and in this, anthropology has a key role to play. This book is about radio, which is not a standard topic for anthropologists. My own research was done under the rubric of Gramsci and Williams and was conceptualized as a study of processes of hegemony and counter-hegemony. This is not only a matter of experience but also of technology, organizations and cultural forms.

In studying this, tape recordings and videos are part of the evidence. My commitment is not to a single research method—ethnography—but to a question about the relationship between radio and political struggles. What is the *field* when studying radio broadcasting? The answer involves challenging the rules of ethnography: no longer to study a "community" in physical space but to follow the people, follow the thing, even follow a metaphor (Marcus 1998). What does it mean to follow the radios? Growing up in Ireland, I played soccer on the streets with sweaters for goal posts and an ear for cars that temporarily stopped the game. There was no referee but our own sense of the rules and fair play. At the start, we would look at each other and ask, are we playing offside?

Alan O'Connor
Toronto, Ontario, Canada
December 2005

Acknowledgments

As usual there is a long list of people to thank. The first must be Luis Ramiro Beltrán, special advisor to UNESCO for Latin America in Quito, Ecuador.

In Bolivia, I had the kind co-operation of Alfonso Gumucio of the Centre for Alternative Media (CIMCA).

Thanks also to Sandra Aliaga (communication researcher, La Paz), Arturo Archondo (Qhana, La Paz), Víctor Baldivieso (FSTMB, La Paz), Leda Nooshia Burwell (Radio Bahá'í, Otavalo, Ecuador), Leónidas Cámara (FSTMB, La Paz), Marco Chicaiza (Radio Otavalo, Ecuador), Enrique Conejo (Office of Indigenous Radio, Quito), Marcelo Córdoba (CIESPAL, Quito), Fernando Dávila (Radio San Gabriel, La Paz), Luis Davila (CEDEP, Quito), Gloria de Vela (CIESPAL, Quito), Ralph Dexter (Radio Bahá'í, Otavalo, Ecuador), Roberto Durette (Radio Pio XII, Llallagua, Bolivia), Argelia Estevez (UNDA-AL, Quito), Dennis García (CEDECO, Quito), Andres Geerts (ALER, Quito), Jorge Gomez (*Chasquihuasi*, Santiago, Chile), Roland Grebe (ERBOL, La Paz), Hugo F. Hannover (Radio Bolivia, Oruro), Attilio Hartmann (UNDA-AL, Quito), Javier Herrán (Cayambe, Ecuador), Rosemary Kane (Radio San Rafael, Cochabamba, Bolivia), Gridvia Kúncar (communication researcher, La Paz), Andy Laughlan (ALER, Quito), Osvaldo León (ALAI, Quito), Clemente Mamani (Radio San Gabriel, La Paz), W. Patricio Muñoz (Escuelas Radiofónicas Populares, Riobamba, Ecuador), Antonio Papue (Shuar Centre, Sucúa, Ecuador), Antonio Peredo (*Aquí*, La Paz), Eduardo Pérez (Radio Fides, La Paz), Fernando Reyes-Matta (ILET, Santiago, Chile), Raquel Salinas (*Chasquihuasi*, Santiago, Chile), Freddy San Miguel (Radio Uncía, Bolivia), Helga Serrano (communication researcher, Quito), Eduardo Tamayo (communication researcher, Quito), Runacunapac Yachana Huasi and the community of Simíatug (Ecuador), Rene Torres (Radio Católica Nacional, Quito), the Union of Campesino Organizations of the North of Cotopaxi (UNOCANC) (Ecuador), Edgardo Vásquez (Radio Continental, La Paz), Ramiro Veizago (La Voz del

Minero, Llallagua, Bolivia), Alan Yoshioka (copy editor and indexer, Toronto) and Eric Mills (proofreader, Toronto).

Research in Ecuador and Bolivia was supported by the Social Sciences and Humanities Research Council of Canada.

Introduction

Anthropologists have not said much about radio. This is surprising because the subaltern populations that are most often the subjects of anthropology books are very likely to listen to radio broadcasts. Rural people turn on the radio very early in the morning for music, local news and the time. Marginalized people in the city listen to the radio because there is often nothing else to do. One often sees a small transistor on a market stall, helping to while away the long gaps between customers. Old buses crawling up the mountains in the Andes play radio stations through loudspeakers. Many stores in small towns have a radio on, and some broadcast the programming into the street as a kind of advertisement that they are open for business. One suspects that for a previous generation of anthropologists this was all an annoyance, like a radio through a thin wall when one is trying to sleep in a cheap hotel. They did not travel all this way to hear something apparently quite similar to commercial radio at home.

There was an assumption that radio broadcasts were part of modern complex societies. The study of radio listeners was left to sociology. One of the first research projects of applied sociology in the United States in the 1930s and 1940s was to study radio broadcasts and audiences (Adorno 2000, Lazarsfeld 1946). There were concerns about the effects of radio propaganda on urban populations. But the boundaries between sociology and anthropology are not so clear. The sociologists quickly discovered that the kind of small groups that interest anthropologists played a crucial role in mediating the radio message. It was called the two-step flow of communication. Informal group leaders were shown to play a crucial role in accepting or rejecting innovations suggested by mass media (Katz and Lazarsfeld 1955). These early researchers on radio audiences also turned to "traditional societies" to learn what they could, as it happens, to improve the broadcasts of the Voice of America during the Cold War (Lerner 1958; Samarajiwa 1987). Yet most anthropologists continued to ignore the radio sets that they heard while doing their field research.

Part of the reason for this is that anthropologists usually did intensive fieldwork in a particular village or community. Radio broadcasts cover much larger areas than most anthropological fieldwork. If ethnography is based on sharing the lives of people in a limited geographical area, radio broadcasts are likely to seem a nuisance, an obstacle to doing fieldwork. Today as anthropologists move away from imagining their work in this way, radio broadcasting becomes a legitimate topic. Radio involves issues of how the community is situated in relationship to the nation (in the case of national broadcasters) and to flows of commodities, migration, even the exchange rate of the local currency against the American dollar (in the case of commercial radio stations). In other words, radio is about what connects the village or community to an economic and political system that affects its life but seems quite beyond everyday experience.

The English cultural studies theorist Raymond Williams invented a term for this need to connect lived experience to economic and political systems: a "knowable community." He argued that no village or community exists apart from these underlying global systems. Every village is internally shaped by the demand for its commodities and work, by the national government that seldom leaves the village alone and by wars that call its people to serve in the army. There is therefore, Williams argues, an urgent need to have a sense of this larger system. Newspapers, novels, films, social studies and more recently television offer such "knowable communities." It is important to understand that Williams is using the term with quite an amount of irony because part of his point is that these larger structures are often puzzling and relatively unknowable.

A few anthropology books written in the past twenty years mention radio[1]—usually at moments when larger structures of power or the economy put everyday life in crisis. At these moments the anthropologist herself may turn to the radio, looking to connect what is happening outside the window with what Williams calls a knowable community. June Nash talks about being in Bolivia during a political crisis. She stayed in her apartment and turned on the radio. "There was nothing to do but sit in the shuttered living room listening to the transistor radio, with the bombs close at hand" (Nash 1979a, 362). Norman E. Whitten describes the arrival of the President of Ecuador in the remote town of Puyo. His speech on development was broadcast live on the local radio station (Whitten 1976, 10). In her famous testimony as the wife of a Bolivian miner, Domitila Barrios de Chungara (1978) also describes the role of the miners' radio stations in the struggles of the mining community against the national government and the army.

Some anthropologists have understood that Williams's work links the ethnographic interest in community with larger economic structures (Marcus and Fischer 1986). Williams not only writes about knowable communities in a global context. He also has a passionate interest in democratic media (Williams 1976). The two interests are related because his whole point is that many aspects of communities are deeply unknowable through lived experience and are only available as statistics, social science, news.[2] In the mining communities of

Bolivia everyone is aware of the world price for tin (Latin American Bureau 1987). For Williams, the relationship between lived experience and capitalism in the countryside is described in the novels of Thomas Hardy. His novels show ordinary speech, work and desire. They describe not a timeless rural countryside but the underlying and varied patterns of change, economic forces, the mechanization of farming and the anonymous connections by which food is delivered from farms to people in the city.

> They reached the feeble light, which came from the smoky lamp of a little railway station; a poor enough terrestrial star, yet in one sense of more importance to Talbothays Dairy and mankind than the celestial ones to which it stood in such humiliating contrast. The cans of new milk were unladen in the rain, Tess getting a little shelter from a neighbouring holly tree. . . .
> . . . "Londoners will drink it at their breakfasts tomorrow, won't they?" she asked. "Strange people that we have never seen? . . . who don't know anything of us, and where it comes from, or think how we two drove miles across the moor tonight in the rain that it might reach 'em in time?"
> —Thomas Hardy, *Tess of the D'Urbervilles*, quoted in Williams (1975, 252–3)

It is not an isolated village culture. It is connected in complex ways with the circulation of commodities: here the marketing of dairy products through the railway system. Hardy is both a participant and an observer of this world.

In one of the rare early anthropological texts that mention radio we can see the American anthropologist Oscar Lewis taking up the stance of what Williams calls an "isolated moral observer"—a stance in sharp contrast to Hardy's. Lewis is writing about the life of a poor family in Mexico City.

> On Mother's Day, I came home singing a song we had been rehearsing in school. "Forgive me, dear Mother, because I can't give you anything but love." My father was at home and he seemed very proud and happy about something.
> "No, son, we can give her something else, because just look at what I bought." I saw a little radio standing on the wardrobe.
> "How nice, *papá*," I said. "Is it for *mamá*?"
> "Yes, son, it's for *mamá* and for you too."
> That's how my father spoke to me then. He had won on his lottery ticket and bought it with the prize money. Afterwards, I came to hate the radio because it caused arguments in the house. My father got angry with my mother for playing it so much. He said that it would get out of order and, "Nobody pays for anything around here except me!" He wanted the radio on only when he was at home. (Lewis 1961, 25)

Here the radio belongs with the forces of modernization, with changing life in the city. And in the life of the family the radio acts as a disruption. It is the cause of arguments between the parents. In a different and more experimental work of anthropology the radio is present as part of a world that is a mix of the traditional and modern. Here the American anthropologist Whitten describes life in a small town in Ecuador, a kind of border town between national life and the hidden but always changing world of the Amazon:

> Nor does he wish to remain in a world of cement, dust, grime, garbage, and foul water, even though he enjoys its fried fish, rice, baked bread, and broiled chicken, its beer and *trago*, its hubbub, movement, marketing activity, movies, radio, and jukeboxes. The food and drink tastes no better to him, though, than fish and meat prepared by María, cooked with hot peppers and salt, and served with steaming plantains and manioc, and the beer is no better than *asua*, still warm and with a wide range of tastes and sensations. He enjoys listening to music on a little Sony radio he has, but he also loves to hear the song of *jilucu*, the whisper of bat wings, the slight, ubiquitous, and familiar sounds of the forest and its fringes. (Whitten 1985, 241)

One can see in the prose a shift from anthropologist as observer to something closer to participant and observer. Whereas for Lewis the radio set is a symbol of modernity that disrupts family life in Mexico City, for Whitten the reality is more complex: "he enjoys listening to music on a little Sony radio." The frontier town is part of his life, and he moves back and forward between it and the forest.

Possibly the most famous statement on radio and social change is not by an anthropologist but by the revolutionary Frantz Fanon. His essay on the Voice of Algeria was written before he became disillusioned by the Algerian middle class, which was taking over the independence movement. His essay is an inspired hymn to the possibilities of radio. Ordinary Algerians, Fanon says, associate the technology with the French occupier. There are many reasons not to listen. But this quickly changes when the national liberation movement starts broadcasting from nearby countries. Then portable receivers and batteries are quickly sold out. Fanon describes listening to the broadcasts as a psychological commitment to national revolution. It even becomes an exercise in guerilla strategy when the French block the signal and listeners have to search the dial for the new frequency. Now to purchase a radio set requires a permit from the French authorities and batteries are unavailable. But with commitment there is a way. This article is in many ways a summary of Fanon's revolutionary passion (Fanon 1965).

Quite different is the contemporary United States research on radio and "development." It is now clear that Daniel Lerner's *The Passing of Traditional Society* was based on data originally collected as listener research for the Voice of America broadcasts to the Middle East (Samarajiwa 1987). Lerner describes with enthusiasm the role of urban media in creating a "revolution of rising expectations." This was part of the contribution of American scholarship to the Cold War. In plain language this means the role of the Voice of America in persuading listeners of the virtues of consumerism and presidential elections. Fanon was not impressed.

This line of research continued in the 1960s with the work of Wilbur Schramm for the United Nations Educational, Scientific and Cultural Organization (UNESCO). After fairly unconvincing arguments about the general need to invest in media for national development, Schramm (1977) switched attention to the virtues of small media. Other researchers focused more exact attention on the potential of radio in specific campaigns to improve nutrition or agriculture. The current wisdom is that such campaigns are of limited use but are still worth some investment (Hornik 1988). Communication consultants who offer their services for such projects seldom mention the real context for poor health and malnutrition: small farmers' lack of land and resources, and the collapse of world prices for many agricultural products.

One of the most important influences in this field in Latin America is the writings of Paulo Freire. In a playful piece of philosophical writing, he argues against agricultural "extension" and for genuine communication between the agronomist-educator and the peasant, whose world must be encountered as a totality.

> It is forgotten that even when campesino areas are touched by urban influences—through the radio, the easiest communication, and by means of the roads that lessen distances—they nevertheless keep certain basic nuclei of their way of being aware. These forms of awareness are different from urban ways, even in their manner of walking, dressing, speaking and eating. This does not mean that they cannot change. It simply means that these changes are not given mechanically. (Freire 1973, 118; translation modified)

This holistic emphasis on communication and respect for another way of life is in practice a moderate position on social change. It has been especially influential with the radicalized parts of the Catholic Church, which is very active in community media in Latin America. What is kept is the emphasis on dialogue, though it is unclear how dialogue is possible in a one-way medium like radio broadcasting. Most of the radio stations discussed in this book attempt to grapple with this problem. What falls away, especially during a period of conservative Church leadership, is Freire's argument that dialogue will not genuinely be possible without fundamental changes in social structure and land ownership.

In the field of media research one finds some attention to local radio stations: small commercial radios in Bolivia (Gwyn 1983) or religious radio stations in Ecuador (Hein 1984). The most famous examples of local stations are the Bolivian radio stations owned and operated by the miners' trade unions (O'Connor 1990, 2004). More recently, the miners have been experimenting with television (Huesca 1997). Media researchers are often extraordinarily upbeat about the possibilities of radio broadcasting. Anthropologists tend to ask more questions. Blanca Muratorio writes that "in a society like Ecuador, where the mass media are directed exclusively at a white or mestizo audience, a Quichua radio station becomes a very powerful form of ideological penetration among the Indian population" (Muratorio 1981, 515–6). Writing about Catholic and Protestant missionaries and their relations with the Achuar people, Anne-Christine Taylor repeatedly mentions radios as desirable consumer goods. "Both organizations promote the sale of short-wave radio sets, since radio transmission is for both an important element of propaganda" (Taylor 1981, 657).

Much writing on the crisis of anthropology strikes me as somewhat irresponsible. Where else can social activists turn for reliable books on the complexity of issues underlying revolutions that increasingly spring like tigers from marginalized parts of the globe? How many people outside the region actually knew anything worth writing down about Nicaragua before 1979 or Chiapas before 1994? When the flood of publications about the Zapatista uprising of 1994 dies down, there is no doubt that some of the best books will be those written by anthropologists who were there long before the television cameras arrived. Another area that would benefit from anthropological knowledge is the many "development projects" launched each year. Independent researchers with an anthropological knowledge of the area assess few of these projects. For example, hundreds of uncritical evaluations were written about radio schools in Latin America from the 1950s to the 1970s. We owe the very few critical reports on these radios to the institutional independence of those writing the evaluation and also to the social science and anthropological background that they bring to the task.[3]

The usual equipment of an anthropologist is a pen and notepad and an endless willingness to listen to people. Some researchers bring a tape recorder or a video camera. The method of anthropology is centered on the idea of community fieldwork: getting to know people. There seems to be a tacit agreement that technology such as a radio receiver disrupts the work of ethnography. Mine is a Panasonic RF-B60 with the essential short-wave band and equipped with an antenna that extends as long as my arm. It is about the size of a paperback book, is quite heavy and rapidly uses up locally manufactured batteries. It is mostly propped up against a pillow in some cheap hotel room. With a small cassette tape recorder I record sample programs through the speaker. My radio receiver is an essential piece of equipment in my ethnographic fieldwork, though I also visit radio stations and talk with people who work there and try to understand the regions to which they broadcast.

To study low-power radio stations you've got to travel, and here you bump into other professional travelers: anthropologists, Catholic priests and sisters, evangelical broadcasters of all kinds, specialists in communication and development projects and the occasional radio technician—certainly the most welcome visitor to any radio station. The most interesting writing on the possibilities of democratic broadcasting comes from unorthodox researchers such as Eric Michaels (1994), who worked with aboriginal pirate television in Australia. More generally, as ethnography about local subaltern populations becomes increasingly problematic, George Marcus (1998) has suggested a series of other strategies: follow the people as they migrate, follow the thing, follow the image or allegory. After ten years, I can now say that I have been following the radio stations.

This book starts with two uses of rural radio in development projects. I argue that the expert knowledge of anthropologists is necessary in order to assess these projects in a broader context. Chapter 2 examines the debate between theories of oral culture and community organizing through one particularly interesting experiment in democratic radio in which indigenous organizations make weekly tapes for an early-morning program broadcast by a Catholic radio station in Bolivia. I then turn to a remote community in Ecuador whose radio station is part of its complex connections with indigenous movements for social change throughout the Americas. Chapter 4 deals with the famous miners' radio stations in Bolivia in relationship to the dominant media. The last chapter looks at a taped news service for Latin American community radio stations, in part a practical response to debates about global communication.

This book does not describe a knowable community in Williams's sense. It is rather more fragmentary, part of a larger mosaic, like fragments of a film. I have written, as I understand it, a modernist rather than a realist text. The reader is left to decide upon debates such as those between the Catholic Church and indigenous values (see the documents translated in the Appendix). The government literacy program that appears in all its ambiguity in Chapter 2 appears again in the discussion of the Latin American news service in Chapter 5. Some of the people who are described in early chapters also reappear in the final chapter. It seems that in presenting local stories for a Latin American audience, the fundamental issues are made clear but some of the ambiguities are left out. This is not to criticize a Latin American news service that was done brilliantly on a tiny budget. The task remains for everyone concerned to imagine the lived experience of those whom Fanon calls "the wretched of the earth" in a knowable community. This will not be easy, and radio stations alone cannot do it.

---

1. Before my initial fieldwork in Bolivia and Ecuador in 1987–89, the most interesting account of indigenous people and radio I read was in Richard Wright's travel book

Cut Stones and Crossroads (1988). He describes in some detail the early morning commercial broadcasts directed to indigenous listeners in the city. This use of commercial radio by indigenous radio producers is also described by Llorens (1991). For an interesting account of radio programs that market traditional medicine in Ecuador, see Miles (1998).

2. For another discussion of the idea of a "knowable community" see O'Connor (1989a). In different ways, the emphasis on situating lived experience in a wider context that is not completely knowable is central to many works of anthropological writing in the 1980s including Mintz (1985), Taussig (1980, 1987), Whitten (1985) and Wolf (1982).

3. A team of German researchers wrote the first major critical evaluation of radio schools in Colombia (Musto 1971). A Mexican researcher affiliated with Stanford University wrote a critical evaluation of similar schools in Mexico (Schmelkes de Sotelo 1973). An official indigenous radio station in Chiapas is sharply criticized for its institutional racism by an independent Mexican-American researcher based in the United States (Vargas 1995). These independent assessments are the exceptions to the rule of "insider" assessments that are normally guaranteed not to rock the boat. For representative examples see Hein (1984, 1988) on Radio Bahá'í, Cornejo (1996) on state-owned "indigenous cultural radio" in Mexico, and the pre-revolutionary Camilio Torres (1971) on Radio Sutatenza in Colombia.

Chapter 1
Radio and Development

It's a busy day at the Cochabamba market. Women in wide skirts, braided hair and round white hats sit on the ground behind small amounts of goods for sale. Others wander between the vendors carrying a woven basket for their purchases. Some women carry a young child in a cloth on their back. A man walks by holding his son's hand. Here in the market you can buy almost anything: sweaters, blankets, pots, pans, plastic buckets, soap, toothpaste. There is also prepared food if you're hungry. This is the third-largest city in Bolivia, and an important agricultural region surrounds it.

That's what the video shows next. People from the surrounding villages head home on foot carrying their purchases. Five men and women walk in file towards the village of Tarata. It's an old theme in development studies: modernization comes from the cities, where education, literacy, mass media and political participation are more advanced (Lerner 1958; Schramm 1977). Modern life comes slowly to villages like Tarata, but perhaps radio can speed things up. That's what the video suggests. After all, if villagers can be persuaded to buy soap and menstrual pads, why couldn't the same techniques be used to sell literacy, latrines and clean drinking water?

A development project attempts to use radio to persuade people in the Cochabamba area to use soybeans in their diet. There are problems of chronic malnutrition among the children. But after several decades of experience in development communication, enthusiasm about the role of mass media has worn thin (Hornik 1988). Too many projects have failed. In this case, cooking demonstrations are organized in the communities because it is now believed that small group meetings are crucial to changing people's behavior. Radio and printed materials (posters and recipe books) are used to complement the cooking demonstrations. Announcements on the radio remind people about the meetings and afterwards reinforce the message.

But there is another innovation. This soybean project was funded by the United States Agency for International Development and carried out from 1978 to 1980. Development agencies were turning to a philosophy known as social marketing. It was the use of advertising and marketing techniques to sell ideas. In this case it was marketing nutrition. Says the video (1980):

> Because the project wanted to develop a viable marketing scheme and not rely on a government give-away approach, sales were handled commercially. The project provided soybeans wholesale to existing tradespeople who in turn were allowed to make a small profit on the retail sales. A price was set which allowed soybeans to compete effectively with other local beans.

When the meetings and cooking demonstrations were finished, radio was used to remind mothers about the soybean message. The project used a Catholic radio school, Radio San Rafael in Cochabamba, and two small commercial stations, Radio Armonía in Cliza and Radio Continental in Punata. There were one-minute advertisements. A boy scores a goal in a soccer match—because he eats soybeans. Using these commercial radio stations is described as more viable than experiments in educational radio that often end when the funding is terminated (O'Sullivan-Ryan and Kaplun 1978). These village radio stations have close links with listeners, who pay for musical requests—popular among courting couples or for birthdays. There are other advantages. "While some stations (such as the one operated by the miners' union) present political viewpoints, the village stations are careful never to criticize the government" (Gwyn 1983, 82).[1]

The Question for Anthropology

I often show this video about the soybean project in Cochabamba to my students in a Canadian university. Then I break the class into small groups and ask them to discuss it. Their responses are different. Some ask where the soybeans come from. Are they surplus production from the United States? But many students feel very uneasy about the exercise. I don't know anything about this culture, they say. Or they feel that they have no right to make a judgment about it. Some argue that if the soybeans help with nutritional problems, then the project is worthwhile even if aspects of it seem questionable. What questions are raised by this project and how should it be evaluated?

In the city of Cochabamba you can take a bus from below the railway station to Cliza. The bus passes through the valley and past many small fields. Cliza is a sleepy town on a Saturday afternoon, but Radio Armonía news can be heard on loudspeakers outside the household appliances shop. Different radio stations can be heard coming from houses, including some that are very poor. I hear one playing Radio Libertad from Cochabamba. There is a wedding in the church. Radio Armonía is now playing rock music in English. The church has loudspeakers on each side of its tower that were used for the wedding but now are blaring out music in Quechua for vendors and about twenty people in the plaza. Radio Armonía plays an ad for beer: "Our tradition is the unity of all Bolivians, our tradition is Taquiña beer." The church loudspeakers echo very effectively in the plaza and through the neighboring streets. The household appliance store has turned its loudspeakers off. At 3 p.m. in the plaza a Catholic nun is running a Girl Guides–type meeting.

The Catholic Church and the radio station—even the USAID soybean project—are interventions in a historical and political field. Even before the 1952 revolution in Bolivia, there was organized peasant political and educational activity in the Cochabamba Valley. The nearby town of Ucureña was the site of the first campesino union in the region. After a great deal of struggle, organized peasants in the area arranged in 1936–37 to rent part of a Church-owned hacienda and thus to eliminate an intermediary exploitative landlord. Needless to say, there was sustained resistance by other local landlords to this dangerous experiment (Dandler 1983). After the 1952 revolution, peasants throughout Bolivia lapsed into sometimes conservative political alliances to defend their newly obtained land. When peasants once again emerged into the political sphere in the 1970s, making new demands on the state, the first armed confrontation between the army and peasants occurred in the Cochabamba Valley (Dunkerley 1984, 210–15). It is surely no coincidence that this area was chosen for the USAID soybean project.[2]

The Ucureña cooperative was both political and cultural. A school was established for campesinos' children, and this building rapidly became a center for political organization. Schoolteachers had a crucial role in helping campesinos to organize and formulate their political demands. Jorge Dandler describes what happened using Eric Wolf's notion of "cultural brokers." The traditional nexus had looked like this:

> Indians—clergy—administrators—hacienda bosses—national and
> provincial authorities (conservatives)

By the 1940s it was replaced by an alternative one:

> campesinos—union leaders—schoolteachers—national allies,
> lawyers, politicians (modernizers)

This model draws attention to the historical role of schoolteachers. It also raises issues of intellectuals and hegemony, the state and political power.[3] There is a clear understanding that education and culture are deeply political matters. In the video on the Cochabamba soybean project, there is a moment when the team arrives in a small village to show a film and discuss issues of nutrition. "By nine o'clock the next morning the demonstration team was setting up its kitchen equipment in the local school. Village people were still arriving and excitement was growing about these outsiders." Two demonstrators wearing uniform orange tops and blue skirts spread a white cloth on a wooden table outside the school. A group of twenty campesino women in wide shirts, some with round white hats, arrive in a tight group. It is another nexus:

> campesino women—project demonstrators—radio stations—USAID project administration—national government

The USAID project is not simply a generous attempt by the United States to aid the rural poor in Bolivia. It uses the techniques of development communication and advertisements on local commercial radio stations to create an alternative political nexus. This continues the USAID strategy in the Cochabamba Valley of using aid in an attempt to replace the formal leaders of the campesino union by non-political "informal leaders" (Dandler 1976).

Radio Bahá'í in Ecuador

On Calle Jaramillo in Otavalo, just up from the Poncho market square, there is a yard with two large wooden gates. On one is written lightly in chalk the letters FICI. The initials stand for the Federation of Indians and Campesinos of Imbabura. All day long and well into the evening, Otavalo Indians in their traditional black dress and hats may be seen entering and leaving. Behind the wooden doors are two busy offices where local people consult with elected FICI officials. In the yard, groups of ten or fifteen Otavalo Indians in traditional dress stand around chatting. The federation receives assistance from at least one non-governmental aid organization. There is a closely linked organization called FRECIDI, the Indigenous Cultural Front of Imbabura, which has its own office, cultural center and library above a popular restaurant on the Poncho market. There is no way to ignore FICI. It organized a protest against increases in bus fares and the racism of the bus operators. The graffiti from this campaign are still visible on the walls of the town. The city of Otavalo celebrates a yearly festival each September called the Fiesta Yamor. It is largely oriented to tourists, but FICI is active on the organizing committee and within the program there is in effect another festival by and for the Otavalo Indians. During the 1987 festival, FICI loudspeakers flooded the Poncho Plaza with folkloric music in Quichua. A painting competition for indigenous children was organized by FICI, with the children using the market stands for their paintbrushes, paints and paper. The organization even videotaped the event. Also during the week of Yamor, FRECIDI organized a three-day exhibition of indigenous films in a local cinema. The films were free and the cinema was filled to overflowing each day. Apart from a handful of young gringos, the audience was entirely made up of Otavalo Indians, many in traditional dress.

Radio Bahá'í is located at the edge of Otavalo, several streets beyond the Poncho Plaza. The signs say, "Radio Bahá'í of Ecuador: The Family Station." The building is a converted middle-class dwelling, now too small for the station's needs, with the usual surrounding fence and gate. In 1987 there were seven staff members, all originally from Otavalo, and the co-ordinator, from outside the country. All staff members are members of the Bahá'í faith, although this is not a condition of employment. They are paid an *ayuda* or assistance rather than a proper salary. There are two producer/announcers who work in Spanish and four who work in the Quichua language. There is no commercial advertising. Many different kinds of music are played but only a small amount of rock. All albums are censored, and items about sex or drugs, or contrary to the Bahá'í faith, are highlighted on the album cover and may not be broadcast. In keeping with the Bahá'í faith, there is no national or international political news. There are no newscast programs as such. The station uses a magazine format with a mixture of music and spoken messages. They are in Quichua from 4 a.m. to 7 a.m. and from 2 p.m. to 7 p.m., and in Spanish from 7 a.m. to 2 p.m. The station also broadcasts educational programming supplied by Radio Netherlands, Spanish National Radio, the United Nations radio service, FUNDAEC (a US-funded organization for the application and teaching of the sciences) and the Friends of Radio Bahá'í. All of this material is in Spanish, and this education programming is simply put on the air. There are generally no organized listening groups or instructors.

The magazine programming contains frequent messages about the Bahá'í faith.

> We present a teaching about the unity of humanity by Bahá'u'lláh, the founding prophet of the Bahá'í faith. . . . The light of men is justice. The intention of justice is to be the unity between peoples, among whom it occupies the highest state and the highest status. The lights of rectitude and necessity challenge and shine from such arms. . . . That was a message of the Bahá'í faith.[4]

The station also accepts written messages to be read over the air as part of its general magazine programming. These messages generally concern matters such as lost goods or family contacts. The station naturally makes a selection. No announcements are accepted about goods for sale. Radio Bahá'í does not allow community groups to make their own programming for broadcast on the station.

As part of its political work, FICI has started to become involved in communication. In the 1980s the organization produced a short-lived newsletter called *Yayarishun*. Bruce Whittington (1985) reports that the organization produced several radio programs for broadcast on the local commercial station, Radio Otavalo.[5] The coordinator of Radio Bahá'í refuses to accept programming from FICI. The reason given is that the Bahá'í faith forbids involvement in politics. Only items relating to "community development" are acceptable. The coordinator explains that a written announcement from FICI about an arts festival would be acceptable. However items that (these are her examples) denounce the president of the republic or argue that the Communist Party is better than the other parties are unacceptable. The federation has approached Radio

Bahá'í in the past and again recently requesting airtime. It has been refused. According to the Bahá'í faith, conflict must be avoided. Social change will occur through the development of each human being to his or her full potential. The result of this policy is quite a lot of hostility between the supporters of FICI and the radio station. For example, during the important bus conflict in 1986, the radio station did not support the users' strike against the bus companies and argued for reconciliation rather than conflict.[6]

In his 1984 article on the Bahá'í radio in Otavalo, Kurt Hein scarcely mentions the religious aspect of the station. His discussion is in terms of democracy in communication. He claims that the location was chosen because of its high number of Bahá'í faithful. How did a small town in a predominantly Catholic country such as Ecuador come to have sufficient numbers of the faithful to justify needing their own radio station? Is it possible that the real purpose of the broadcasts is to win indigenous people to the Bahá'í faith? What are the effects of this on the community? Protestant missionaries are already quite active among the Otavalo Indians, and there is potential for further divisions within the indigenous community between the supporters of the radio station and activists within the indigenous federation.[7]

Hein describes the town of Otavalo and its surrounding valleys as a rural traditional society whose way of life is being eroded by twentieth-century technology and values. Traditional weaving is becoming mechanized. Influenced by outside media, young people leave for jobs in the cities. The purpose of Radio Bahá'í is to strengthen the indigenous culture by providing broadcasting by local people, much of it in Quichua. The station encourages traditional music and provides education, social services and development education. It also provides a forum for announcements and messages between local people. Local people participate in the station. One day I asked some campesinos in the Cayambe area why country people buy a radio. Among the reasons are to know the time and to listen to music. People listen to Radio Bahá'í for the music, they said. Another time, during the bus drivers' strike, I drove a large group of people from an indigenous community to the hospital. Along the way I talked about radio, and when we got to the hospital they asked me for a sheet of paper to write a personal message to be broadcast on Radio Bahá'í. As well as this kind of participation, Hein concludes, there is some degree of self-management of the project by the indigenous radio staff.[8]

As with the Bolivian soybean project, the language here is one of traditional ways of life and urban social change. From an anthropological perspective, the reality is much more complex. The Otavalo Indians are weavers and traders, as well as farmers, and have skillfully adapted themselves to social and economic pressures. They have kept their traditional dress, language, music and artisanal activities and are the most prosperous group of indigenous people in Ecuador. Hein draws on a study of Otavalo migrants to Quito. But this study actually shows that even when they migrate to the city, Otavalo Indians are relatively successful in keeping their link to their culture. Far from being threatened by modern technology, this ethnic group is flexibly adapting to the dynamics of a dependent capitalist economy in Ecuador, changing certainly, but always keeping what is important for their cultural survival (Weinstock 1973). Of all the ethnic groups in Ecuador, the Otavalo Indians least need outside help in maintaining their culture. The real problem, though, is that Hein offers no further analysis of the political economy of the area in which Radio Bahá'í is located. It is clear from other studies that the cause of migration is not the introduction of Western values or even technology, but lack of land and agricultural inputs and the near impossibility of making a living by producing for the domestic market.

The separation between anthropology and development communication allows questionable projects to proceed and to be reported in uncritical terms. For Richard J. Gwyn and the USAID project team, the problem in Bolivia's Cochabamba area is due to poor nutrition and this can adequately be addressed by a campaign to persuade women to use soybeans in their cooking. According to Hein in Ecuador, the problem of the Otavalo Indians is that mass media and modern life are eroding their "traditional" way of life. From an anthropological point of view this kind of analysis is completely inadequate. So too are the solutions proposed using rural radio stations in Bolivia and Radio Bahá'í in Otavalo.[9]

---

1. One of the most important statements of this approach to development communication is by Robert Hornik (1988). His underlying assumption is that social change, such as the redistribution of land, credit and agricultural inputs, is not possible. His emphasis is on information. He completely ignores Latin American approaches to communication such as the critiques by Freire (1973) of agricultural extension and by Beltran (1980) of vertical communication. Hornik's work is closely linked with USAID. He describes with enthusiasm a project, funded by the United States during the Somoza dictatorship in Nicaragua, to teach mathematics by radio. The satellite instructional television experiment he describes in India is now recognized to have been a failure. He is wholly enthusiastic about the ACPO radio schools in Colombia, apparently unaware of critical studies that show the schools had a conservative effect on their students and did not support a movement for land reform. Because of the separation of the fields of anthropology and development communication, experts such as Hornik are almost never challenged about the implications of their interventions, and they continue to act as consultants to governments around the world.

2. In January 1974 an army massacre of between 80 and 200 campesinos in the Cochabamba Valley effectively ended the alliance between the national government and campesinos against the miners and other workers.

3. These issues are most clearly formulated in the writings of Antonio Gramsci (1971).

4. Recorded live from Radio Bahá'í on 6 September 1987.

5. In an interview in September 1987, however, the manager of the station could not recall this.

6. The bus strike is documented by Colloredo-Mansfeld (1999, 73–5). For general background on Ecuador see Corkill and Cubitt (1988).

7. On the activities of Protestant missionaries in Ecuador, see Goffin (1994), Muratorio (1981), Taylor (1981) and Hvalkof and Aaby (1981). It is also necessary to ask critical questions about the role of the Catholic Church, which is very active in the field of radio in Latin America. There are radical bishops who support a "preferential option for the poor" and significant movements for social change. But this is in the context of an increasingly conservative church leadership. Even radical bishops can be conservative on issues of sexuality, birth control and abortion. For useful background see Löwy (1996). Each radio project sponsored by the Catholic Church needs to be independently evaluated.

8. Hein's own involvement in the Bahá'í faith is not stated in his article. In fact he worked at the station and later as director of a Bahá'í radio station in the United States.

9. For a recent overview of issues in the Otavalo region of Ecuador, see Korovkin (1997) and more generally Ibarra (1987). For the Cochabamba area in Bolivia, there is an excellent study by Lagos (1994) as well as the historical work of Dandler (1983).

Chapter 2
Oral Culture or Social Organization

Men, women and children are helping to sort the potato crop, which is spread out under the sun on a large patio. This is a community of indigenous campesinos a few hours from the city of Latacunga in Ecuador. These are Spanish-speaking people but three members of the community still speak Quichua. A boy of ten or twelve years proudly shows me his notebook in which he is learning to write. Small children are playing under the supervision of two or three young women. This community center has a building with a large classroom, a dormitory and outdoor cooking area, the office of the campesino union and a simple radio studio. This studio has a worktable, two tape recorders, one microphone and about fifty cassettes. There is a photomural on the wall about a *minga*—a collective workday for the community. More construction is in progress. Yesterday one of the volunteers had an accident when a concrete block fell on his hand. He was taken to hospital, but the community still had his severed finger in a small cardboard box covered with a white cloth. A youth comes to warn the union president that two soldiers are approaching in a vehicle. There is obvious alarm in the community, even though it is likely that the soldiers have come only to check up on yesterday's accident.

The Ecuadorian anthropologist Marcelo Córdoba is visiting the Union of Campesino Organizations of the North of Cotopaxi to discuss several matters, including its participation in the national literacy campaign. Marcelo is well known and liked in the community, and is quickly engaged in a deep discussion with the Union president. The Union does not necessarily support the new government but views its literacy campaign as a service that members should be willing to use. The national coordinators of the campaign are experienced and the teaching materials seem to be good. However, the Union feels that it is important to orient the literacy workers to the history of their struggles, to basic ideas of political economy (concepts of work, value and exploitation) and their culture, as well as making sure that they can teach literacy. There will be 217 participants from the campesino union and about twenty-five outside literacy workers. The present Union leaders are mainly graduates of the Popular Radio Schools of Ecuador (ERPE), an earlier generation of radio school that broadcast from Riobamba.[1] Over the past two decades the value of these radio schools has been questioned, but here the literacy skills they taught have been valuable for the community leaders.

The Union of Campesino Organizations of the North of Cotopaxi is an independent campesino "organization of the second degree." As its name suggests, it is a union made up of community organizations; the latter are "organizations of the first degree," and the distinction in degree is important. Within the Union there is a specialized committee for communications that prepares a weekly program for broadcast on Radio Latacunga. Popular reporters—people's reporters—inform listeners about *mingas* and meetings. Women have their own organizations within the Union and play an important role in the radio programs. Democratic discussion takes place within the Union—but is not heard on the radio program itself. People I talked with were doubtful about a European-style radio open to all opinions. They pointed out that there are people who would like to destroy their union. Radio has its uses, but a radio station of its own is not high on the organization's list of priorities.

Radio Latacunga is owned by the local bishop, and this puts limits on what can be broadcast. The station did transmit a radio drama about a campaign for land. It also took a position on the issue of an increase in bus fares. People working at the radio station gradually push it in a more radical direction. However, there are limits. The bishop recently forbade access to supporters of a political party—he is a supporter of the Christian Democrats. The radio station is part of one of the most interesting experiments in democratic radio in Latin America. Seven or eight campesino unions are equipped with simple recording equipment and make a weekly program to be broadcast as part of the early morning show on Radio Latacunga. Several international non-governmental institutions are involved in this development project, which is coordinated by Marcelo's current employer, the International Center for Advanced Study in Communications for Latin America (CIESPAL), an international institution with its own office building in Quito.

The Union of Campesino Organizations of the North of Cotopaxi participates in this experiment in *radio popular*. Marcelo visits to discuss their activities. One of the issues today is a political conflict between the union and the Church-sponsored Indigenous Movement of Cotopaxi (MIC). Internal elections are coming up within the MIC, and Marcelo and the union president discuss the possibility of a more radical committee being elected. This would reduce conflict between the two organizations. Another item they discuss is a development project sponsored by Habitat. The Union carefully selects the aid organizations with which it works in order not to be swamped by numerous projects and lose control of its destiny. A strong campesino union works with only two or three carefully selected aid partners and ensures democratic control over the process. As Marcelo and I talk, we notice that it is now past noon. The community members invite us to share their midday meal: potatoes, rice, lentils and a hard-boiled egg.

The Voice of the People

The person who invented the idea of simple radio *cabinas* (studios) in the countryside is Javier Herrán.[2] He is a likable and hardworking Catholic priest. During the bus strike he continually stops his jeep in the mountains to give country people a lift. At their destination they make a gesture to pay him something, but he waves them away with a laconic "*siga no más*" (just carry on). The idea of the radio *cabinas* came from his observation of popular but commercial radio stations such as La 11 de Noviembre and Radio Saquisilí: what was popular on them was news, messages and dedications.

During lunch we talk about indigenous culture, which for Herrán is above all an oral culture. He thinks that Radio Bahá'í is not bad. The people at the station respect the culture. A lot of people listen to it. Indigenous culture has its strong and its weak points, though. For example, news is mixed up with personal opinion, and the foundation of these opinions needs to be questioned. We are travelling in his jeep with a campesino leader and his wife in the back seat. I turn to involve them in the conversation, and we talk about why country people buy a radio.

We visit three communities during the afternoon. The first has a new church building, and a house is being constructed for Catholic sisters. A Peace Corps volunteer is working on the plumbing. We give a lift to a Catholic nun. At the next stop there is a community center (a religious lesson is in progress in the meeting room), a health clinic and a community shop. A kitchen is being constructed with *minga* labor by eight or ten men, and a few women are also helping. The nun who came with us asks if there have been any visitors with propaganda (by which she is referring to Protestant missionaries). Our last stop is at a community that is the focus of a potable water project. These are all communities rather than organizations of the second level. However, Herrán is also the founder of the Indigenous Movement of Cotopaxi (MIC).

The slogan of Radio Latacunga is that it is "the voice of the people on the march." But the emphasis is mainly on the spontaneous expression of indigenous culture. Herrán played an important role in the development of the radio station. He tells the story in this way:

> Radio Latacunga went on the air in December 1981. From then the mothers' centres, the youth groups, the Church assemblies, the campesino community houses (*casas campesinas*), but fundamentally the Indigenous Movement of Cotopaxi (MIC) have been the principal beneficiaries of this rich experience of popular communication. . . .
> From the start the necessity was clear for two types of programming:

> one in Spanish for urban centres and Spanish-speaking campesinos;
> and the other in Quichua for the indigenous communities. The first in
> the conventional manner of a commercial radio; and the second
> "Ricchari" totally unedited and spontaneous in the protagonists' own
> voice. This is the true Radio Latacunga. (Herrán 1987, 25)

This gives great emphasis to two factors. First, the focus is entirely on community groups such as mothers' centers and youth groups rather than on organizations of the second level. Second, Herrán insists on spontaneous oral expression. This is in accord with the origins of the Latacunga radio *cabinas* project. The initial tapes for broadcast were not done with scripts. The first short tapes were informative and emphasized who, what, when and where, says Herrán. They developed into tapes with community news, personal messages, ads for community shops, and so on. There was drama in which campesinos took the roles of political boss, landowner, drunk and campesino, and discussed an issue. Herrán thought of these radio plays in three acts on the model of classical Greek drama. There were interviews, stories and jokes.

The International Center for Advanced Study in Communications for Latin America (CIESPAL) had a small project near Latacunga in Salcedo using printed materials and slides. The center got involved in the Radio Latacunga project because Herrán needed both finances and assistance. Their first training courses were technical, but later they did training on radio forms. Herrán argues that this introduced ideas from outside that disrupted oral expression and introduced organizations into the work. He points out that there is no Quichua word for organization. It also introduced what he sees as a political ideology and led to difficulties with Radio Latacunga. A self-sufficient radio station would not have problems with the bishop. Herrán is critical of the bureaucracy of CIESPAL, which had two full-time organizers on the project, and Radio Latacunga, with its staff of eighteen persons. In his opinion the project should have been kept simple.[3]

The activist priest is starting a new radio project around Radio Mensaje in Tabacundo. Unlike Latacunga, the communities here are dispersed rather than organized around market towns. Organizations of the second level are virtually extensions of political parties, says Herrán, and impossible to work with. He plans two master studios: one in the Salesian buildings in Cayambe and one in Tabacundo. They will send signals directly to the transmitter. Each studio will be fed by *cabinas* equipped with very simple equipment: a cheap turntable, a small mixer, cassette deck and microphone. The station itself will have only two full-time employees, who will play music, give the time and read material. A half-hour of national and international news will be rebroadcast from Radio Católica Nacional in Quito. Otherwise all production will be from the *cabinas*, which will result in a simple, inexpensive service. He also plans a small charge for personal messages and musical dedications. I ask him about access to land,

and he replies that this area has the best land in the country in campesino hands, individually owned and worked. It is richer than the Latacunga area.

Problems of Indigenous Radio in Latin America

Many of the difficulties of *radio popular* are evident in the documents of the Seminar-Workshops on Radio in Indigenous Regions of Latin America, held annually in the late 1980s (Instituto Indigenista Interamericano 1986, 1987, 1988). For a start, indigenous broadcasters are underrepresented because of difficulties in travelling to the meeting, especially when it is outside their own country. Those presenting tend to be Catholic clergy who specialize in radio, project organizers and anthropologists. The presentations early in the week are so vaguely worded that it is impossible for the participants to come to any conclusion: generalizations abound about giving indigenous people a voice, about being heard, about overcoming marginalization. Such language easily covers projects that are quite questionable in their organization and intentions. A representative of the Instituto Nacional Indigenista (INI), an organization of the Mexican state, questions the religious motives of many projects sponsored by the Catholic Church, whereas everyone seems too polite to ask about the role of the Mexican state itself in sponsoring indigenous radio through the INI.[4]

One of the initial desires expressed at the second Seminar-Workshop, held in Mexico in 1987, is for exchanges and mutual aid between radio stations. As is often the case, what is needed most are exchanges of news features. Other perennial issues are technical training, funding and legal guarantees of the stations' independence. It quickly becomes obvious that radio broadcasting is *always* part of a larger project, whether a development project, popular education or a popular organization. For example, in Peru organized campesinos may wish to use their participation in the radio to denounce human rights violations. Not all these can actually be broadcast. The station needs to verify the facts, and the organization is in physical danger for speaking out. There are important differences between programs from a low-power community radio (which could have close relationships with those who participate) and programs broadcast at a regional or even national level. Some religious radios are mostly about indoctrinating their listeners; others have a more dialogical aspect. State-owned indigenous radio is unsatisfactory, though some state networks are more responsive than in the past.

As the week-long seminar progresses, the language becomes more specific. Three sets of issues are identified. (1) forms of radio programs that are appropriate for the listeners; (2) radio organizations that belong to indigenous peoples; (3) the relationship between the radio and ethno-development, the improvement of life and protection of human rights. Broadcasting in indigenous languages, indigenous music, free personal announcements for listeners—all of these can easily be found in radio stations that operate in a hierarchical manner. A policy to rescue indigenous languages and culture does not necessarily extend to empowering the indigenous organizations to define their own problems, solutions and political projects. The Instituto Indigenista Interamericano does not have funds to start indigenous radio stations, nor can it pressure Latin American governments to grant indigenous organizations their own independent stations. The limitations become obvious.

Finally, the evaluation of radio broadcasting must include close study of recorded programs. There is little precedent for this, and it is unclear what the method should be to evaluate them. Should indigenous languages be spoken "correctly" or as in daily usage? Should the station play only indigenous music or also music that the listeners actually enjoy? Should "radio magazine" formats be used? Should interviews be long and indirect (following actual speech) or edited to be short and to the point? Is the intention to make *indigenous radio* or simply to broadcast for indigenous people? How are these matters decided?[5]

Issues of Organization and Radio Form

Is there such a thing as objective social science? We already know that Javier Herrán has a poor opinion of CIESPAL. And Marcelo Córdoba, the anthropologist most closely associated with the Latacunga radio project, will argue that successful radio studios are those closely linked to democratic campesino unions and federations. How then do we read a study organized by CIESPAL that compares in detail three radio studios in Ecuador, one weak, one moderately strong, and one strong?

The weak studio at the time of the study is the original studio located in the parish of Zumbahua. Formerly part of a second-level organization, the studio is operated by one person. It gives only news of the weekly market and one local community. The organization, informally controlled by the local priest, has no links to other second-level campesino organizations except to the Church-sponsored MIC. It has no democratic assembly, though the Church has revived local traditional indigenous political structures. An incredible twenty-seven development institutions are working in this area, with the result that there is no overall planning and no possibility of control by the local people. This radio studio was started on the initiative of Herrán himself. It was once very active, but now its tapes frequently have to be edited at Radio Latacunga before being broadcast.

The moderately strong studio belongs to the Union of Campesino Organizations of the North of Cotopaxi (UNOCANC), described at the start of this chapter. This organization consists of sixteen communities. It was started in part from the campesinos' own struggles and in part by a development institution. The Cotopaxi union is in the process of building a democratic structure, including the participation of women. A monthly assembly gets moderate participation. The union has initiated contacts with other second-level organizations. It works with four development institutions and has some control over the general direction of development projects. The radio studio was started on the initiative of CIESPAL. It has a good number of reporters but not from all the communities. The tape consists mainly of information about organizational meetings and events. Controversial topics are avoided. Discussions take place within the organization at meetings rather than in public on the radio.

The strong radio studio belongs to the Union of Campesino Organizations of Mulalillo (UNOCAM). This union of seventeen communities was born from the campesinos' own struggles. It has a strong democratic structure with well-attended assemblies every fifteen days. It has made official contacts with other second-level organizations. It works with three development institutions and is in control of the development plans. The radio studio was started on the initiative of the organization itself. It even sought finance for the equipment on its own initiative. All the communities participate in the making of the tapes. Apparently it has been decided not to have designated reporters. The programs include matters beyond the organization, and the radio style is attractive and colloquial.

The CIESPAL-based report concludes that an autonomous, democratically organized campesino federation has the resources and energy to support a well-organized radio *cabina*. It will produce high-quality tapes with contributions from the campesino unions grouped together in its federation. The explanation for the present poor quality of tapes from Zumbahua is that the studio has never been independent of informal control by the Church. Its organization is rudimentary and it lacks broadly based participation. It takes a culturalist or indigenist political line, rejecting political alliances that stress social class, patterns of land ownership, access to credit and agricultural supports, and the role of the state (de Vela 1988).

In response, Herrán stresses the relationship between the radio and the oral culture of the Quichua-speaking campesinos. Above all he values spontaneous expression. He does not deny that he gave training in appropriate radio forms but, for example, his dramatic dialogues were devised by the group and rehearsed a few times, not scripted. He contrasts the liveliness of oral expression with the halting reading of a script by people who do not have much schooling or opportunity to read. He criticizes the displacement of a tradition of locality and community by what he sees as foreign ideas such as "organization."

For its part, CIESPAL agrees that at first it lacked experience with appropriate radio forms. But Marcelo argues that this theory of oral culture is mistaken. It ignores the fact that indigenous culture is always in a process of change and that there is no original or oral culture to be simply preserved. CIESPAL argues that after some initial difficulties, campesino radio producers become skilled at elaborate radio forms and successfully adapt traditional music and spoken forms (riddles, jokes) to the radio. The institution defends its training in a radio magazine format that it calls the *Radio Revista Informativa*. This scripted, planned form provides a suitable format for communicating the life of the campesino organization. Here is an example of a typical item from a tape prepared at the UNOCANC studio. This program included the voices of both women and men, mixing local music with organizational announcements.

> On the 22nd and 23rd of May a course will be given on forestry man-agement, where all the leaders of UNOCANC who are affiliated will be present. Equally, there will be *compañero* representatives of the institutions and one *compañero* who is in charge of forestry manage-ment at national level. Also present will be *compañeros* from Tungurahua Province, and a representative of the Forestry Committee and a technician. It is arranged, the first day, the 22nd, to cover the following points. First in the morning the *compañeros* of the institu-tions will deal with deforestation and management. And then we'll all go to the cooperative of C—— to continue there with the practical part. The *compañeros* are glad about this [practice] because it is very important.[6]

This scripted item is closely related to the internal organizational life of UNOCANC. But once in place, this organization and form of radio can tackle issues well beyond the organization itself. This second tape is in the voices of two women.

> Communication: A communication to all the campesino organiz–ations of the Province of Cotopaxi. There is a problem in the com-munity of Paniatug, which has come out to defend the *compañeros*, the poor *compañeros* of Paniatug. It's that Señor C—— wishes to seize the land, and we as campesinos have to help. Today it's for them, tomorrow for us. This communication is from all the leaders of the organization.[7]

This actual call to action against the seizure of community land (there was a protest march of more than a thousand campesinos) is embedded in a tape that is not cold or foreign. The same half-hour *Radio Revista Informativa* has a humor-ous story, music, a riddle game, laughter. And in moments of crisis the program can quickly make radical demands on the economic and political system.[8]

---

1. The Popular Radio Schools of Ecuador (ERPE) in Riobamba were experiencing internal problems at this time and campesino access was suspended, though there were popular reporters for local news in Quichua and Spanish. This is in part related to the general questioning of the radio school model (see O'Connor 1993) and in part because of local political differences.

2. The idea of two-way radio communication was discussed in the 1930s by Bertolt Brecht (1983). Hoxeng (1977) describes the use of tape recorders at Radio Mensaje in Ecuador. Mario Kaplun's cassette-forums also included two-way communication (ALER 1983). From 1969 his organization, Latin American Radio Service (SERPAL), produced

taped dramas for use in group discussions. The groups recorded their responses and these were circulated. SERPAL's funding was withdrawn by the Catholic Church and the organization collapsed (interview with Andres Geets, 17 May 1989).

3. Its status as an intergovernmental institution does place limits and pressures on CIESPAL. For example, its directory of institutions with *comunicación popular* education projects in Ecuador (CIESPAL 1988) includes many conservative projects of the government along with others that are much more progressive.

4. Vargas (1995) found institutional racism at an average INI radio station in Mexico. For a typical example of uncritical research on the INI stations in Mexico, see Cornejo (1996). The demand by the Zapatistas that these stations should be handed over to indigenous organizations is well known (Marcos 1995, 158).

5. The Third Seminar-Workshop (held in La Paz in 1988) made distinctions between indigenous radio and radio broadcasting in indigenous languages for commercial or political purposes. It identified the need to meet the needs of indigenous youth (who otherwise turn to commercial radio stations). An office for indigenous radio was established in Quito, Ecuador, for three years. This important initiative is documented in Instituto Indigenista Interamericano (1986, 1987, 1988).

6. Broadcast on Radio Latacunga on 24 May 1989. Recorded and translated by the author.

7. *Radio Revista Informativa—Curso Taller de Nuevos Reporteros*, Saquisilí, 2–5 August 1988. Tape supplied by Marcelo Córdoba. Translation by author.

8. See also the forceful statement on behalf of the Comité de Cabinas de Grabación de Cotopaxi by Dioselina Iza (1986).

Chapter 3
The Mouth of the Wolf

Perhaps because my copy of the video is several generations old, it makes the long shots of the mountains look like oil paintings in deep green and brown pigments. The wind howls through the microphone. Simíatug in Quichua means the mouth of the wolf. A heavy truck covered in a tarp rocks its way along a narrow unpaved road. In medium shots it looks picturesque against the mountains in the background. Meanwhile two other figures on donkeys make their way down the mountain. They are shown against the sparse undergrowth covering the bare mountain. The mestizo truck driver speaks. He was born in Simíatug but no longer lives there. He is a merchant and brings in goods, above all for the market on Wednesday. Two boys stand on top of the truck, seen against the mountains and the weaving road. The two figures on the donkey are arriving from a different direction. The truck driver explains the difference between mestizo and indigenous people in our country: they are badly brought up, there are things they don't understand. The wolf howls through the microphone. And there nestled in the hills is the town of Simíatug, with its white houses arranged along the curving entrance road into the central square.

The houses look abandoned, the paint peeling from the walls. The camera shows a small sign that reads: For Sale. Another voice tells us that there only remain thirty or forty white families. Many houses are padlocked. Of those who remain, many are women. One operates a soft-drink bottling operation in her home. She is a widow and inherited the business from her husband. She fills bottles with a funnel from a large plastic tub, and a boy adds the gas and caps the bottle. She explains that the natives purchase clear soft drinks and cola as medicine. Things have changed in the town. Now the natives say that they are the same as us, she explains. The camera shows an open fire and there is food on the table. In another house with peeling paint, a woman explains that now she lives more apart. Before, things were cheaper. Now when the campesinos have meetings to decide things, she says, we are not taken into account.

For the first time, we hear a spokesperson for the indigenous people. He wears glasses with black frames and sits among a large group of silent people wearing ponchos and hats. The camera shows close-ups of work-worn hands. He explains that things could not continue as before. We were at their orders, we couldn't say a word. The indigenous people are now organized, he says. This is the only indigenous voice we hear in the video. But now the truck finally arrives in Simíatug, along the narrow streets and past the large church building, painted white. A woman's voice tells about the market that was organized in the town because the native people had products to sell such as cereals and animals. The streets are now crowded with people in black and red ponchos, women carrying babies. Indigenous men purposefully walk into shops. Some drink and play billiards in the towns bars. Goods are exchanged in the marketplace and money counted. Then people begin to leave by foot or on donkeys. Unsold pots are packed up. People impatiently load heavy sacks into the truck, which then slowly pulls out of town, leaving the streets to an occasional dog or a group of small boys. On the soundtrack a woman sings a hymn and prays that the town will not be completely destroyed. In her ways of seeing things, the town is in control of the natives. They're going to kill us, she says. A very large group of indigenous people in dark and bright ponchos go into the cemetery, following a coffin painted deep blue. The mists swirl again over the town, and the wolf cries into the microphone.

The Radio Technician in Quito

Andy Laughlan is a radio technician, a young British volunteer who has been working with the Latin American Association for Educational Radio (ALER) for three years. During this period he has been in Simíatug for a total of about six months. He knows more about Radio Runacunapac Yachana Huasi, the local indigenous station, than anyone at ALER. He sketches the history of exploitation of the indigenous people by mestizos who controlled access to the market. They would block the road and force indigenous producers to sell or threaten to destroy their goods. The Catholic Church also extracted taxes in kind and then sold the goods to the mestizos. Gradual solidarity by the indigenous people developed into a base organization that is highly democratic. Following the example of the radio schools in Riobamba, the Simíatug federation—Runacunapac Yachana Huasi (the house of the people's knowledge)—started the radio station with help from the Shuar Federation. The "radio quality" of the programs may not be very high by outside standards, but they are listened to avidly because they are by and about the local people themselves.

The Church cannot claim credit for the radio station. The Simíatug federation is suspicious of the Church because of its role in the past. The only connection is that the first radio director was a lay pastor from ERPE, the radio school in Riobamba. There have been problems in recent years with aid. The federation is very suspicious of those from whom it accepts help. At present there are technical services, training and technical advice from ALER. After several years of building this relationship, the radio station has become affiliated with ALER. Oxfam America is also involved through a local organization. This raises problems because all of these local organizations are informally affiliated with a political party and are not disinterested. They're attempting to gain support for their party. A trainer in broadcasting was chosen by a development organization—to the disappointment of a local person and the Simíatug federation. Tape recorders, cassettes, and salaries for popular reporters also were provided. There was an attempted imposition of a Western hierarchical structure with a director paid more than the others, but Runacunapac Yachana Huasi successfully resisted this. Oxfam Britain is providing help with bilingual school materials. The ministry of education is also providing some schools, and the new priest (who abolished the Church tax on goods) is also interested in starting schools. The influx of large amounts of money (which in rumours gets inflated) creates the possibility of tension and suspicion within the Simíatug organization. This is a serious problem because its strength has always been in its unity.

Notebooks and Memories

The best day to visit is probably Wednesday because it is market day. You can get a ride in a truck or a four-wheel-drive car, which is needed on the unpaved road. Access is very difficult for outsiders. The weather is cold and the food is mostly rice and potatoes.

Though my field notebooks from the time mention nothing of the episode, I remember well how I rented a small truck in Quito with no driver's license: they said my passport would do. How I was warned to put the anti-theft lock on every time I left the vehicle. These trucks are very popular, they said. How I struggled to find the exit from Quito to the highway. At one point I turned the wrong direction into a one-way street and a policeman stopped me. It's just that I don't know the streets very well, I explained. I'm trying to find the way to Ambato. But you have to pay for your sins, he said quietly. It took me a few moments to realize that a small banknote would solve the problem. Overnight in Ambato I stayed at a small hotel because it had a secure yard for parking, and the next morning I had difficulty maneuvering the car out. I drove slowly over the worsening roads to Simíatug, enjoying the beautiful scenery and the mists over the mountains. I was also wondering what to say when I got there. I had a letter of introduction from a priest active in community media, but he had warned me that the coordinator of the radio station was very separatist. And you, said the priest, looking me up and down, are very white.

I finally arrived in the early afternoon and attracted quite a bit of attention. I must have told somebody that I had come to visit the radio station because soon I was sitting in the studio with a group of radio workers. The studio was in the

Runacunapac Yachana Huasi (the house of the people's knowledge) beside the church on the main square. It was a courtyard with meeting rooms surrounding a volleyball court. The station was well equipped and had a tape library, some records and a portable recorder. I started to talk about my research on popular media and found that they were very well informed about popular movements and radio stations. There were eight workers at the radio, and one of them had visited Nicaragua as part of a four-person delegation from Ecuador. They had very firm political views. The discussion was cut short by Eusebio Sigcha, the co-ordinator of the station, who explained that I would first have to get permission from the community. There would be a meeting that evening. He said that they were busy with meetings all day and asked why I hadn't phoned to say I was coming.

After supper there was a meeting in a hall with many people sitting around the wall: mainly community teachers, radio workers and organizational officials. It was mostly men, though three women sat together. The group wanted to hear about daily life in Canada. Are there poor people? Indigenous people? Why is Canada richer than Latin America? What products are produced in Canadian factories? Cars? Armaments? What popular organizations exist? What are the problems of the indigenous people? I said that they had land claims, problems with forestry and oil companies. Ah, the same as here, someone responded. What do things cost in Canada? To what social class do I belong? Who paid for my trip to Latin America? What is the purpose of my research? They seemed to think that writing a book or article and the exchange of information between Latin American countries is a worthwhile purpose. After a pause, somebody told me that it was a good meeting—I was free to talk with people in the community tomorrow, which was market day. Later Eusebio told me that people often come to visit the station. He said they wouldn't have talked to me without the letter of introduction. He told me about an American academic who had visited recently, and grudgingly he said that I seemed different. Still, he said, all these people come to visit and we never see any result. It's not worth the time.

At the community meeting it was obvious that these were very well informed people with a lively interest in politics and world affairs such as the student demonstrations in China. They were also very well informed about other indigenous groups in Ecuador. But they were interested in everyday things like my glasses, my short-wave radio receiver (how much did it cost?) and my height (5'10", or 180 cm). That night there was an informal fiesta in the street with brass bands and dancing. There was one couple in traditional dress and much passing around of drinks (and cigarettes for the musicians). Many women sat around wrapped in red blankets. A few danced. An older man gave me a shot of homemade liquor and people were friendly. Somebody said that I danced like a girl. Eusebio came over and carefully explained that the community doesn't party like this all the time: this is a special month for festivities.

I was given a bunk in the community house (after some joking about lodging me with someone who was obviously very poor and who grinned at this), which was a hostel for people visiting the federation from the outlying areas. In the morning, friendly women in the communal kitchen made sure I got a boiled egg and a plate of rice. I recorded the evening show on the radio before the fiesta and the morning broadcast, which started at 4 a.m. Afterwards I hung around talking with people. Eusebio was surprised that I was leaving the same day, but when he found I was heading back to Quito asked if he and one other person could come with me. He and a teenaged boy were soon ready. We said our farewells and were soon crawling down the mountain along the unpaved road.

At first Eusebio was reserved and slightly hostile. The community could really do with a small truck like this. It would be very useful. I explained that it was rented for a few days and I couldn't afford anything like this in Canada. He wanted to drive but I thought that was unreasonable and didn't respond. We talked about politics for a bit. He thought that class struggle and issues of culture were equally important. The land is central to our culture, he said, working the land. He said that the film *Boca de Lobo* is a record of the past when the organization was just starting and the mestizos controlled everything. He said that the film was sympathetic to the indigenous people. I thought the film is really racist, but I didn't want to ask him if he'd actually seen it. He carefully checked to see if I had learned the fundamental lesson from my visit: the strength of the community was in its collective organization. I didn't want to take advantage of our trip to pump him for more information. We stopped for something to eat and got chicken and chips. The teenaged boy had been quiet throughout. They were going to Quito for a meeting to plan activities and protests for the 500th anniversary of the arrival of Columbus in the Americas.[1] I dropped them off in Quito outside the meeting place. There were warm farewells from both. Eusebio even told me to remember that I was welcome back in Simíatug.

The Radio Station in the Field of Power

The canton of Simíatug is located in a remote part of Bolívar Province, which includes several climates: sierra and subtropical. The radio station is owned by the Runacunapac Yachana Huasi indigenous campesino federation. Although campesinos are the most important part of its audience, the station defines itself as a community radio—for everyone in its listening area. The spokespersons for the station insist that it is impossible to understand its purpose without understanding the social and economic history of the area. Several haciendas in the area had long expropriated the indigenous lands. With land reform in the early 1960s some land passed into campesino hands, but exploitation by mestizos continued with the support of the local priest. Campesinos were forced to sell their produce at very low prices. In Simíatug they were exploited by mestizo merchants for the goods they bought and by the owners of *cantinas* or bars (Pachano 1986). Health services were inadequate. Schoolteachers taught only in Spanish, were woefully ineffective and never stayed long.

A campesino protest against these conditions in 1962 was suppressed by the army. From the early 1970s unofficial groupings of communities refused to sell their goods except at market prices. They also began to struggle for official recognition of local teachers who spoke Quichua. The federation was formed in 1976. It represents twenty-three communities or about sixty per cent of the population of the area. It has organized services for health, education and communication, a co-operative store and flour mill. The federation is a member of ECUARUNARI (the indigenous movement of Ecuador) and has economic and political contacts with other indigenous organizations.

In 1981 an abandoned transmitter was obtained from another indigenous group (the Shuar Federation) which has operated a radio for some time, and broadcasting started the next year. It is interesting to note the precedents of the earlier generation of radio schools in Ecuador. Some of the present leaders of Runacunapac Yachana Huasi were students of a radio school (ERPE, broadcasting from Riobamba). However the Simíatug community came to reject the radio school model and instead opted for a station closely linked to the progress of the organization. Programs are bilingual in Spanish and Quichua. The radio did not create the organization, concludes Garcia, "it was the organization that created the radio to consolidate its organizational process, as one more tool in the struggle to change the power structure in the areas of Simíatug" (Garcia 1985, 39).

Many people come to the radio station to record or broadcast for the radio. One effect of this has been to strengthen the musical groups of the area. However, there are no facilities for taping outside the centre of Simíatug. The station broadcasts from 4:00 to 8:00 a.m. and in the evening from 5:00 p.m. There were problems at first because one person tended to dominate the radio. A new radio team was organized from 1988, and since then programs have been lively and well organized. Spanish is used somewhat more than Quichua. There is the usual problem of sources for national and international news, though *Punto de Vista* (from Quito) and *Tercer Mundo* (from Santiago, Chile—see Chapter 5) tape services are used. The daily news is taken from Radio Latacunga and other stations and rebroadcast with additional commentary from Radio Runacunapac Yachana Huasi. There is a high degree of political knowledge and sophistication.

Programming recorded on the 30th and 31st of May 1989 gives a sense of the intention of the radio. As it happened, on the first day there is a meeting in Simíatug of persons from many of the communities that make up the organization. The program on Tuesday from 4:45 p.m. is one of music and greetings from those at the meeting to their family and community at home. The program is bilingual but Spanish is used much more than Quichua. Music is autochthonous and folk music. There are also several batches of communications concerning meetings in various communities and also the six-monthly evaluation of the radio by the communities. A woman broadcaster reads one group of messages. All the other voices are male. The next day, the radio starts just after 4 a.m. with a program called *Wake Up with God*. After this there is a program about reforestation. The last item is from the morning news show, which concentrates on the national stoppage by bus owners who are demanding a 100 per cent increase in fares and the right to import new vehicles without customs duties.

Radio Transcript Excerpts

Tape 0 (side 1)

ANNOUNCER: [In Quichua]

GUEST: [In Spanish] Thanks, *compañeros*. I'd like first of all to salute everyone in the community of S——, above all my family and friends. I hope you're well. You're there and I'm here for the meetings, and so I can't work there as usual in the afternoon. After communicating with my family and friends, I'd like to communicate urgently to the community of S—— that this Sunday, the fourth of June, a general meeting of all of the community members will open. Already this year we have been working on the matriculations that have not been done because of the educator. We communicate urgently to the president of the parents' association, or whatever secretary, that this Sunday therefore will be a meeting, the fourth of June, to carry out the matriculations. I'd also like to communicate with the community of S——. I've no more to say.

ANNOUNCER: [In Spanish] You as a community educator, do you teach them bilingually or in our own language, so that the communities maintain our own language and the *compañeros* of S—— have to strengthen our own language and our people's heritage?

GUEST: [In Spanish] Well, in respect of this, you know well, *compañeros*, that in some communities we give little importance to the Quichua language. In the S—— community, in truth, there are two standards. There are some who desire to learn in both languages and some who don't desire. In our case no. One could discuss in the meetings concerning this problem, from years back, that we should use both languages. Our own language is the Quichua language.

Tape 1 (side 1)

COMMENTATOR: There are some who simply want to convert people to Catholicism. For them, to be Christian is, how can I say it, to be inferior. Not to protest, to conform completely. In other terms: peace. For them: don't respond, don't be badly raised. When everything is controlled, for them that's peace. When anybody protests, that isn't peace. But we say on the contrary that peace is the fruit of justice, because without justice there wouldn't be peace, *compañeros*.

Tape 1 (side 2)

ANNOUNCER: Well, *compañeros*, here we are again for you all. What's up, *compañeros*? Our sincere greetings to the leaders of the communities, also to the women's groups and to everyone belonging to the organi- zation that continues to fight each day. In this program, *My Land*, we'll talk about problems of reforestation in our communities. Welcome. [*Music*]

That's how it is. And as you know, *compañeros*, we are already at the end of May. In some communities they are already working on reforestation, in others perhaps no. We'd like to remind you of pre- vious programs when we were with you conversing about forestation: the problems that exist. This time we'll talk about some of the prob- lems and doubts that some of the community leaders have, also about the problem of wind and the problem of poor soil. Let's talk about the wind, *compañeros*, because we are already in the month of May, the start of summer when, as you know, we have so many problems with the wind. In past times these problems didn't exist. That's why we have to look for the causes. [*Music*]

That's the point. In the quarterly meetings of the communities we have talked about how we've got to this forestry problem. Think about the wind. Think about our bare land: desert. For this reason the quarterly assemblies in the communities have had the idea to have communal workdays to plant trees. Some communities have done this but there have been problems because small delicate trees have sometimes not survived. We'll talk more about this. But now to make us happy here is some music. In this case, musicians from one of the communities perform it.

Tape 3 (side 1)

ANNOUNCER: The theme of our news program this morning is the national stoppage by the bus owners. It continues and people are suffering much in the cities. We'll talk to someone who came from Ambato to Simíatug this morning. Did anything happen?
INTERVIEWEE: Nothing happened at all. It was all quiet.
ANNOUNCER: Is the city of Ambato fortunate? Is it the same as usual?
INTERVIEWEE: No, it isn't.
ANNOUNCER: What's missing?
INTERVIEWEE: Well, because of the stoppage you can't just travel anywhere.
ANNOUNCER: What did the driver say about coming here?
INTERVIEWEE: He didn't say anything.
ANNOUNCER: How much are they charging?

INTERVIEWEE: Right now they're charging 350.
ANNOUNCER: What was it before the stoppage?
INTERVIEWEE: Just 300.[2]
ANNOUNCER: Did he explain the difference?
INTERVIEWEE: No. And I didn't ask.

The emergence of a social movement

The fundamental discovery here was not the radio station in Simíatug, important though it was. The real discovery was Eusebio: his reluctance to talk with outsiders and his insistence on the importance of organization. And the meeting in Quito, where several non-governmental organizations had their headquarters because Ecuador was regarded as safe. A "safe" place means one that is heavily policed; in the Andes it means a country where indigenous peoples are kept firmly controlled. This is no longer the case in Ecuador (Field 1991; Whitten, Whitten and Chango 1997; Zamosc 1994). Early in the twenty-first century, indigenous organizations are militant and highly organized. They clearly have been building their organization for decades. That's what the meeting in Quito was about. Who knows, for example, what role the quiet teenaged boy who came in the truck with Eusebio and me might be playing in today's militant indigenous movements?

We need look no further than the video *Boca de Lobo* to understand the suspicion of outsiders and their representations of the problems of Simíatug. Where the video shows a ghost town inhabited only by widows and children, the reality is a meeting place of a federation of indigenous communities that has been gradually organizing for decades. The filmmakers show the graveyard, but not the federation meeting hall and the radio station. Where the video describes the remote town through the voice of the truck driver, we discover from the radio tapes that the private transit operators not only exploit indigenous people but also are organizing a stoppage to demand increased fares. Without its own vehicle, the indigenous organization is wholly dependent on bus and truck drivers from Ambato for access to the outside world. The main motif of the video is drawn from the winds in the mountains, naturalized as the cry of the wolf. Yet from the radio tapes we discover that problems of wind and soil erosion derive from deforestation, the extreme poverty and mestizos' exploitation of the indigenous people in previous generations.

---

1. Hale (1994) includes a discussion of the international planning meeting that took place in Quito in 1990. For the Ecuadorian indigenous uprising of May 1990, see Field (1991) and Zamosc (1994).

2. In May 1989 the exchange rate was 520 sucres to one US dollar.

Chapter 4
Radio Voices and Knowable Communities

There is no school today because the teachers, poorly paid, are again on strike against the government. A neatly dressed young girl shows us around her small mining town. Her mother is a nurse. The girl points out the tiny one-room row houses provided for the miners. They are about the size of a small bedroom and have no water or toilet. Water comes from a tap outside, shared toilets and showers are at the end of the street. Just outside the door of each dwelling is a concrete platform for a cooking fire. Several streets are filled with rows of these tiny white-painted dwellings, now silent and empty because of the global collapse of tin prices and the government's policy of privatizing the industry. I have such mixed feelings: horror at these living conditions, and yet dismay that the miners and their families are now being pushed out of their communities and into the shanty towns that surround every Latin American city. Two families used to live in each of these, she explains. Then she brings us along the street to a medium-sized white building, now without a roof. That was the prison, she pointed, where the workers were locked up.

Although high in the Bolivian Andes, this was no isolated community. Its very existence was linked with the global circulation of tin and other metals. At the big mines, the ore swung overhead in large buckets suspended on cables. Linked also with an international marketing organization that until 1985 smoothened the rise and fall of tin prices on the global commodity markets (Latin American Bureau 1987). The teachers were on strike to protest against a government policy of privatization that was also being imposed on the mines. Hence the day off school and also the ghostly row of miners' houses.

This is the challenge for anthropology: to situate lived experience in global structures of economics and power. It would be impossible to imagine the Bolivian miners as an isolated community. Ethnographic description must make connections between their experience and the global market for tin, which was needed, for example, in the early twentieth century to supply soldiers at war with canned food. Raymond Williams explores such issues through the cultural forms in which writers and filmmakers have tried to imagine their relationship to global forces that are not knowable from direct experience. Williams's use of the phrase "knowable community" is in this sense ironic because he is aware that many of the global connections are deeply unknowable. Economic and political theory and statistical information attempt to bridge the gap between lived experience and a complex reality, yet uncertainty remains. Above all, there is uncertainty about the future. Williams's idea is often best understood by anthropologists. George Marcus and Michael Fischer write about the problem of how to represent "the embedding of richly described local cultural worlds in larger impersonal systems of political economy" (1986, 77). They suggest that this problem has become more difficult because of the increasing complexity of global systems.

Sugar is another global commodity. In a work of historical anthropology, Sidney Mintz (1985) shows the complex and changing culture of sugar and its production. One can begin to glimpse connections between plantation slavery in the Caribbean and working class families in England drinking heavily sweetened mugs of tea. Even the internal structure of these cultures and communities is affected by global linkages. The industrialized production of sugar was shaped by, and in turn shaped, the culture of the English working class. By the early nineteenth century, impoverished working families, with more women working and therefore unable to prepare as many hot meals, turned increasingly to a diet of tea with sugar, bread and jam.

Meanwhile, in France the ethnographic work of surrealist Michel Leiris goes in almost the opposite direction. Based on careful attention to everyday detail and avoiding literary artifice in his ethnographic work, he creates a form of writing that moves at the same time inward to the self and outward to the phenomena he seeks to describe. His 1934 work *L'Afrique fantôme* outraged anthropologist Marcel Griaule because of its honest description of a research expedition to Africa and the global traffic in cultural artifacts.[1] Ethnography is a complex matter.

The Voice of the Mines

The video opens with a fairly rapid montage of images: scenes inside several different miners' radio stations, scenes of the busy streets in the mining towns high in the Bolivian Andes, several long shots of rows of housing, three miners emerging from the mine and two entering for the next shift, a housewife preparing a meal, a close-up of the radio receiver on the kitchen table. Two miners eat their meal from enamel dishes and listen to the news on their union radio station. The video, by Alfonso Gumucio and Eduardo Barrios, was made in 1984—just before the crisis in the mining industry—and documents a time when there were twenty radio stations owned and operated by the miners' unions. The first miners' station was started in the 1940s and simply played music for the workers. It was tolerated by the owners until it started to broadcast news. Two miners listen carefully to Radio Vanguardia. "This form of communication is outside official or commercial control. It is a dialogue that disturbs, an activity that prevents exploitation and reinforces solidarity," says the voice-over. Photos and children's drawings show military incursions into the mining centers, and the soundtrack is a tape of the stations linked up to organize resistance to the army (Gumucio 1982; trans. in O'Connor 2004). This is the radio's finest moment.

In ordinary times, the video shows, miners travel on underground trains deep in the mine. The galleries are narrow and rubber boots splash in water. Miners with lamps on their helmets pause to leave an offering to the statue of the Tío (Taussig 1980; Nash 1979b). A miner and his assistant struggle with a heavy drill, making holes in the rock face for explosives. It is exhausting work and they are soaked by water. At this time Bolivia was the world's third largest producer of tin. Yet the miners live in poverty and develop silicosis from breathing dust inside the mines. The average life expectancy is less than forty years. Old equipment and lack of investment in the mines made Bolivian tin the most expensive in the world to produce by the 1980s (Latin American Bureau 1987).

There is a radio reporter in the street from the Catholic station Pio XII. A spokesman warns that people in the mining center should use water carefully because very little is available. A miner in the main plaza of Llallagua/Siglo XX speaks into the microphone and expresses his hope that the planned miners' university in the town will become a reality. There is a shot inside the company store. The shelves, reaching up to the ceiling, are almost empty. Then an outside shot of a widow and her young boy shoveling through the mountain of tailings looking for fragments with tin ore. Then a slow pan over the mountains in the distance. The miners have a soccer team—playing a game brought to Latin America by English sailors (Galeano 1998). There is a cinema that shows mainly second-rate movies from the United States. And they have the radio to listen to.[2]

The miners are always aware of the price of tin on the London Metal Exchange. The collapse of the market price in 1985 had enormous effects on their lives. The reasons for the crisis are complex. Bolivian tin is expensive to mine in part because of decades of underinvestment in the mines, owned by the state since the 1952 revolution. But there is also the political refusal of the developed world to pay reasonable prices for commodities—from coffee and bananas to minerals and other raw materials. The immediate cause of the collapse was hostile action by the Reagan administration of the United States towards the International Tin Council, whose function was to smooth out extreme fluctuations in the global prices of tin. This is needed to allow for long-term investment decisions in mining. In Bolivia itself, a political decision was taken to close down much of the mining industry and transfer resources to agribusiness. This seemed a deliberate decision to destroy the political power of the miners, who have historically been the leaders of organized Bolivian workers (Dunkerley and Morales 1986).

La Hora del País

The most famous of the miners' radio stations is La Voz del Minero, broadcasting at 500 watts on 1370 kHz on the medium band and also on short-wave. The first time I visited the mining center of Llallagua/Siglo XX in 1987, the station was not broadcasting. When I returned a year later, it was back on air. *Amanecer Minero* is an early morning program of music and local notices. The presenter is male and has a polished, professional radio manner. The program has music interspersed with announcements about a general meeting of an organization of miners' wives, a meeting to discuss an alternative plan for the future of a mine, a soccer game today and a youth soccer tournament this weekend, and frequent announcements of the time. A statement is read from university students about a local strike. Brass band music is used as "flashes" to introduce announcements. The important announcements about meetings today and the sports events are read several times in the course of the program. One of the songs is the current hit by Los Kjarkas, "Chuquiaga Marka," about the city of La Paz.

The type of literary modernism that interested Raymond Williams would attempt to bring together the historical movements of collectivities and interior life, memories, dreams. Not in a mood of defeat, but looking to possible futures. Listening to the tape with Los Kjarkas brings back vivid memories of the plazas and steep narrow streets of La Paz. Young people flirting in the Plaza Murrillo, the indigenous Plaza San Francisco, the stalls of the nearby market, a huge demonstration in the streets. In August 1987, riot police with submachine guns and tear gas canisters at all official buildings. I heard that song so many times from radios spilling out onto the streets, a kind of unofficial anthem of the Aymara city.[3]

Radio Fides was founded in 1939 and is one of the most influential radio stations in Bolivia. It occupies a large modern building near the center of La Paz. There are plans to increase its power so that it can be heard throughout Bolivia and even beyond its borders. Owned by the Jesuit Order, it has the status of a private institution within the Catholic Church. The radio station determines its own positions and is not an official voice of the Church in Bolivia. Because of restrictions imposed by the government on the Bolivian media during the 1970s, Radio Fides gained the reputation of being the *vanguardia informativa del país*, playing an important role in the formation of public opinion. For this the army intervened in the station in November 1979. Radio Fides, along with other progressive media, was destroyed during the 1980 coup by General García Meza.

The director of Radio Fides is appointed by the Jesuit Order and is responsible for its programming. The station has a strong and effective direction in the 1980s: conservative and pro-business. The intention of Radio Fides is to strengthen the traditionally weak Bolivian state. Yet the radio station does not seem at first to be conservative. It can point to participation by listeners, some of whom visit it. There are popular programs that include phone-in participation. Two programs, *La Calle* and *Ampliado Vecinal*, include outside broadcasts from the poor neighbourhoods of La Paz. This participation does not necessarily conflict with the goal of supporting the Bolivian state if most of the participation takes the form of petitions to central and local authorities to perform their legal functions.

The station director himself conducts the national morning news program *La Hora del País*. This program has a wide listenership, especially among the Bolivian middle class. This flagship program is "a national endeavor to integrate—in an hour or more—the country with the pulse of the news, through our forty-two correspondents and retransmission by many stations" (Radio Fides 1989). The program links up with a network of Catholic radio stations throughout Bolivia to provide news from many different cities and towns. The pace is fairly rapid, and to this is added the pulse of numerous advertisements. The content of *La Hora del País* matches very closely the news of the daily paper *Presencia,* which is associated with the Catholic Church. A comprehensive survey of alternative media in Bolivia describes Radio Fides as part of the dominant media, although only partly so (CINCO 1987). It does not count Radio Fides among alternative media: important changes have taken place since its resistance to the 1980 coup.

The news program *La Hora del País* is repeatedly interrupted by advertisements for medicines, Canon office machines, Nissan cars, Marlboro cigarettes, a La Paz hotel and banks, and by public health and tax notices. All of these imply a middle-class business listener. The ninety minutes of *La Hora del País* for 19 July 1988 contain forty-seven advertisements: approximately twenty-two and a half minutes, or a quarter of the program. A news broadcast that is so pulsed by the rhythm of advertising jingles seems to have given up part of its meaning. Even the motif of the program (the hour) has been sold to an advertising sponsor at the start of the program. The main announcer, the anchor in La Paz, is called by name. The person of Eduardo is associated with the time, which he gives with precision to the second: the official time for all of Bolivia. The regional correspondents are dispersed in space: they give the local temperature. "Hello Eduardo, hello Bolivia, hello friends, listeners to *La Hora del País.*" What is the function of this repetition throughout the program? This repetition creates the illusion of stability and continuity. You may imagine leaving the brisk morning of La Paz for the warm breakfast time of the Bolivian Amazon or the South. But you will always return, in good time, to the acoustic image, to the grainy voice of Eduardo *en estudio central.*

I interviewed Eduardo Pérez in his office at Radio Fides, which occupies a large four-story building, almost a fortress, in downtown La Paz. He is a Jesuit, about forty years old and wears a three-piece suit. Previously on the extreme left, he was exiled in the 1970s and is now pro-business. He is also communication adviser and consultant to the mayor of La Paz, a member of the conservative ADN party. Pérez has headed Radio Fides since the return of democracy in the early 1980s. I am admitted into his large executive office. He is extremely businesslike. "You have ten minutes," he says. He answers my questions, but I feel as if I am reminding him of the radical he used to be. It is the only time in two years of research that I get kicked out of someone's office.

Back in the mining town of Llallagua/Siglo XX, tune your radio in the early morning to La Voz del Minero, the most famous of the miners' radio stations. Its own midday news bulletin covers events in the town of Llallagua and the historic Siglo XX mine. But the station does not have the resources to deal with national and international news. For this it relies on Radio Fides from La Paz. Starting at seven in the morning, it picks up *La Hora del País* and rebroadcasts it. Except the announcers block out the ads with music. When I listened in 1988, this was done roughly. Sometimes a bit of an ad got through before they covered it with music. Sometimes they were a second or two late in getting back to the news.

In La Paz, Pérez sees no problems with the advertisements on Radio Fides. First, this is a plural society: he doesn't drink alcohol or take soft drinks, but others do. Second, they are necessary: this is a commercial radio station, he said. Third, they are like oxygen.

The first miners' radio stations emerged in Bolivia in response to the control of the mass media by the mining oligarchy, including a short-lived company radio station. The miners' own station, La Voz del Minero, started broadcasting from Siglo XX mining district in the late 1940s and early 1950s. Funded by a small weekly contribution from each miner, it was of relatively low power and had rudimentary programming of music, the time and union news. This changed in the early 1960s when a Catholic station was set up in the town of Llallagua to combat communism in the mining centers. A radio war developed between Radio Pio XII and La Voz del Minero, and in order to compete the miners had to hire professional broadcasters from the cities.[4]

Along with the other miners' radio stations, La Voz del Minero has been attacked by the military on several occasions. In 1978 it was almost completely destroyed by the army. In July 1980 it formed part of a "Network for Democracy" opposed to the military coup of García Meza until it was again forced to close. In 1989, because of the government policy of dismantling the state-owned mining sector, there were only 430 unionized workers left at the Siglo XX mine to support the radio station. The union building has a cinema that is rented out to a private operator. Part of the income is used to finance the station. However the continuation of La Voz del Minero on the Bolivian airwaves in the late 1980s seems mainly due to the initiative of five radio workers. They are paid the equivalent of US$19 per month. There is some advertising for local stores, but the union would not permit other types of advertising even if it were available. This is not a commercial radio station.

The station operates within the lines established by the miners' union. It is supervised directly by the union's secretary of culture. There is no committee to guarantee the radio's independence, as exists at some other miners' stations. This has not been a problem in recent years, although in 1976–78 changes in union leadership did result in changes in radio personnel. There has been participation for some time now by campesinos' and housewives' organizations, although they do not speak directly or produce their own programs. For news they have two reporters. One covers the Siglo XX mine organizations: the union, the housewives' organization, the union of "relocated" miners and the campesino federation. The other reporter covers the town of Llallagua: the university, the mayor, various organizations. For national news they use *La Hora del País* from Radio Fides. For international news they use the papers *Presencia* and *Hoy*, and listen to the international news on short wave radio. The radical weekly paper *Aquí* is also an important source.

When again I visited La Voz del Minero in 1989 it was still transmitting the morning news program from Radio Fides. This is the point of Williams's discussion of knowable communities. The miners' community high in the Bolivian Andes has its local traditions but has also been internally shaped by the global market for tin and other minerals and also by the pressures from national politics—in turn affected by the policies of international organizations such as the World Bank. A community of resistance has been created by the organized miners over decades of struggle, and these communities are now in part being dismantled. Global and national tendencies in the market and in political decision-making are vital parts of the future of the mining communities. La Voz del Minero, instead of simply blocking out the advertising in *La Hora del País* with music, now substituted its own commentary on the Radio Fides newscast. My memory of this is the delight I felt at this ingenious strategy. From a powerful and conservative organization in La Paz came news items that were often of vital interest for the mining communities. But their own tiny radio station blocked the commercial messages and instead improvised a commentary on what had just been heard. "Well, no we don't agree with the analysis just presented by Radio Fides," a confident and professional voice would say. Against the framing voice of the *estudio central* from the capital city, a different political voice said firmly, "We see it differently."

La Voz del Minero, 21 July 1989, 7 a.m.

[Music fades under] From the largest mining district of Bolivia, transmits Radio La Voz del Minero, the vanguard of liberation, in Siglo XX, Potosí, Bolivia.

[fades up] . . . friends, listeners to *La Hora del País*, a very good morning.

As is logical and understandable, all the political parties wish to negotiate with the other democratic parties that have parliamentary representation, to establish and guarantee the next constitutional period. This appears beneficial and agreeable. And the voices that are raised to object to one party joining with another seem to us to be totally anti-democratic and in reality, destructive. And as always, they matter a button.

The political parties of the center and the extreme right are negotiating. In the cabinet there will be two ministers for each ministry, so those of us who do not even have ten per cent political experience can learn instead of criticizing each other and arrive at more logical positions and in this way build the future of the republic. The question is the response of the left to this. Democracy is clean, but all together. As for the past, let the dead bury their dead.

[electronic noise] Good morning, friends, listeners. We heard the commentary of the columnist from Radio Fides, who agrees with the point of view of Licenciado Jaime Paz Zamora that it is necessary to forget the past. We do not share this opinion. We think that all men, all the parties, are a product of an objective reality, of a historic process that they have been called upon to live.

General Bánzer, in effect, is a dictator. He has very recently repented but no one can guarantee what will be his future conduct. He is the wolf hidden now under a democratic appearance.

What to think of Gonzalo Sánchez de Lozada? We all know and feel what has been the consequence of his obsequiousness to the International Monetary Fund and the consequence of him defending his economic interests. He is today enjoying the fruits of a deal, with his partner, to reach an agreement that is completely against the interest of the Mining Corporation of Bolivia. We think that the MNR [National Revolutionary Movement] and concretely Gonzalo Sánchez de Lozada not only represent the interests of imperialism in our country but have economic interests to maintain themselves in power.

What to say about Jaime Paz Zamora? A party that says it is leftist revolutionary, with its heroes, the heroes of the 15th of April, etc.? These heroes never fought for a concession with the oligarchy. Even so, Jaime Paz has done this. He has permitted the Eradication Law against dangerous substances, for the eradication of the coca. He has supported with his parliamentary vote the presence of North American troops in the country.

We can say, we the workers, that this is not a government of national salvation. Effectively no.

We return to Radio Fides for a summary of the news from La Paz.

---

1. Price and Jamin (1988). See also Richard and Sally Price (1992).

2. On the mining company's introduction of television sets see Barrios de Chungara (1978) and on attempts to set up alternative television stations in Bolivia, Huesca (1997).

3. For a very interesting description of relocated miners in the city, see Gill (2000).

4. For the history of the miners' radios, sources in Spanish are Gumucio (1982), Gumucio and Cajías (1989), Kúncar (1989), López (1985), Lozada and Kúncar (1983), Rivadeneira (1982) and Schmucler and Encinas (1982); in English, Gumucio (2000), Huesca (1995) and O'Connor (2004).

Chapter 5
Beyond the Local

It's a typical day at the Chasquihuasi organization in Santiago, Chile. Five members of the staff are busy putting together the next edition of *Tercer Mundo* radio news tapes. The cassettes are sent by registered mail to community radio stations throughout Latin America. After months of visiting radio stations in the Andes, I've come to Chile to discuss my research project with some Latin American communications researchers, including well-known researcher-activist Raquel Salinas. Her son, who is disabled, spends part of the day here and is noisily demanding attention. He is part of the organization and his job is to put the stamps on the envelopes before the tapes are brought to the post office. The most important part of this institution is its informal structure. In spite of all the activity and rush, there is an overall air of calm and purpose.

Raquel takes a break from preparing a report and her work at Chasquihuasi to discuss my ideas. She is immediately interested in the work of Raymond Williams. Radio as a cultural form is something she's thought about. It is shaped within institutions and it is worthwhile thinking about the formation of radio workers (that is, their background and education). Complex forms of communication are undoubtedly part of struggles over meaning—what Williams calls hegemony—and this has political effects. Raquel asks me to send her a copy of Williams's theoretical book *Marxism and Literature* (1977). I'm concerned about sending her a book with a title like this in a country that is then still far from democratic, but she says not to worry.

Raquel's formation is as a communication investigator working for international organizations such as UNESCO, and as an academic, but above all as a Third World activist, an activist for the New World Information Order. She has worked as an assessor for various alternative news agencies (Salinas 1984). From this work she concluded that such big expensive projects were never going to solve the problem because they don't reach the poorest people. Such alternative news agencies don't get to Radio Latacunga or La Voz del Minero. Clearly determined to help with my research, Raquel goes on to sketch the other people involved in producing the *Tercer Mundo* tape service. Equally responsible is Jorge Gomez, a Colombian who spent seven years working for the Catholic radio network UNDA-AL. This gave him a practical understanding of the problems of poor urban, indigenous and campesino sectors. Especially these sectors lack information in general and practical knowledge. For their news bulletins, most popular radio stations read the headlines from the (dominant) daily newspapers. The popular movements' lack of information about Latin America was discussed at various meetings, including UNDA-AL, which could not deal with it because it had to attend to its own internal problems.

At this point Raquel and Jorge developed a personal relationship and began working together. "Chasquihuasi" is Quechua for "house of the messenger" and the name was chosen as an indigenous/Andino/minority name that would be acceptable in all of Latin America. The intention of the *Tercer Mundo* tape service is to overcome barriers impeding access by popular media to circuits of international news. They believe that they reach a different audience than other news services—an audience of *rural* radio listeners. The technology is not "new" but uses cassettes sent by mail to be broadcast on radio stations. The service is a union of Raquel's studies of the circulation of international news and Jorge's practical experience in broadcasting and organization.

Also part of the team is Mario Villalobos, a grassroots communicator with no formal training in communications but a background as a union leader. He has practical experience in union bulletins. Alexis Parez is a student and sound technician. He is responsible for choosing the music between each item on the tape. Sandra Contreras is the secretary and accountant. Carmen Gloria Espinoza is in charge of documentation and archives. The organization is registered as a private limited company—the only structure legally available. Raquel and Jorge would like the staff to take over the organization as a workers' co-operative, because they have plans to start a school for disabled children in which Raquel's son could participate.[1]

The content of *Tercer Mundo* is of a new type. The idea of "development news" in Latin America never had much meat on its bones. In Africa it clearly means support for government projects. The Latin American media theorist Fernando Reyes Matta (1986) stresses oppositional news. While Catholic radio news was committed to the popular sectors, it did not have international news. What Raquel and Jorge have tried to create is a news bulletin that will be educational. This has two objectives. First, to put local experiences in a global context; for example, to explain why there are poor countries and what in the world situation causes there to be displaced miners in Bolivia. And to explain this in such a way that a person with little education really can understand world economics. Second, to stimulate the organized action of base groups to help themselves to attain basic needs.

It is important that it is not just news of micro situations. The macro and the micro of the story have to be put together. Just as Raquel's big picture and Jorge's practical radio experience at Radio Latacunga and organizational experience at UNDA-AL combine in a family-like organization to produce *Tercer Mundo*. The form of an *informativa* is midway between news and investigative documentary. Pure news is only information—it doesn't educate. *Tercer Mundo* is explicitly not impartial, its values and arguments are not hidden. Why news and not radio drama? Because most listeners—educated and non-educated—do listen to the news. It is also an easier form to manage than elaborate documentary investigations. Furthermore, the form is non-academic. It has the impact of news. Many points can be made in one program. It favors a global vision and there is some continuity from week to week. The half-hour program can be broadcast or it can be used in sections as part of the radio stations' own news programming. Stations can also cut stories they don't like for political or religious reasons.

Mario, the journalist, makes a selection of about fifty stories from different sources: Inter Press Service and bulletins and tapes received, including some from member stations. (The taped reports are retold in the voices of *Tercer Mundo*'s professional newsreaders.) Then Raquel and Jorge meet with Mario to make the final selection. We're the gatekeepers, she jokes, using a term from communication studies. Alexis selects the music to be used between each item. The scripts are recorded in the studio by professional broadcasters (one male and one female). This ensures good sound even if a less-than-perfect radio transmitter in the mountains broadcasts the cassette. The tapes are sent by registered mail (the largest single expense) to radio stations every two weeks. In the beginning it was mostly the UNDA-AL and ALER radio stations (mainly Catholic) that subscribed, but its distribution has now expanded beyond this. The only government radio using the service is Radio Nacional de Panamá.[2]

Radio as Cultural Form

We should think about radio as a complex *cultural form*. The term is from Raymond Williams, and one example is the "knowable community" form of the English novel. What he suggests is not a communication model of transmitter, message and receiver, but radio as a cultural form. His own analysis of British television in the early 1970s is complex. He describes television as a *flow* of programming, rather than something with a beginning and an end like a book. He builds up a picture of this flow of images and sounds as the applied technology of a way of life that he calls *mobile privativism*. This way of life is acutely mobile (car traffic, population movements) but increasingly privatized relative to the working class community of the early twentieth century (Williams 1974).

It is well known that experiments by aboriginal people with video cameras result in tapes that are quite different from standard television. In particular their images of landscape—long slow pans in which nothing seems to happen—correspond to their own sense of place and memory (Michaels 1994). In a similar way, it would be surprising if popular radio in Latin America followed the models of broadcast talk for Britain and the United States. Paddy Scannell's (1991) discussion based on British radio includes key elements that are not universal: public speaking for a domestic audience. The idea of "public speaking" is highly variable in an anthropological perspective. As is the idea of a "domestic" audience—differences in family structure are a perennial topic for anthropologists. *Public institutional talk* is an idea that looks considerably more complex in ethnographic research. Erving Goffman (1981) begins to suggest this with the idea that radio broadcasting consists of the production of a flow of talk—a kind of drama in which the announcer continually risks losing face by making errors of fact or pronunciation or unintended sexual innuendo. But clearly this assumes North American commercial radio and a *presentation of self* that is not universal. The script from Bolivian miners' radio (Chapter 4) is a kind of political commentary that implies commitment to workers' struggles rather than a universal speech community. The material from Simíatug is somewhat closer to "taking the word" at a community meeting.

The *unfinished* aspect of popular radio as a cultural form may remind us of Williams on the English novel. The difficulty, in other words, is to connect knowable communities with global contexts. Or to situate knowable communities in layers of historical time. Put simply, most community radio stations in Latin America have great difficulty with events beyond their own locality. How is the news—regional, national and international—to be found and presented? This, of course, was the subject of an impassioned debate in the 1970s and 1980s, more or less between the developed nations and the Third World.

The New World Information and Communication Order

French film director Bertrand Tavernier, whose work includes *Around Midnight* and *A Month in the Country*, arrives at a meeting at the Writers' Society in Paris. European filmmakers are concerned about access to television for their work. Tavernier has just been to see French president Mitterrand about this, but has returned with no guarantees. He enters a majestic room at the Writers' Society and takes his seat at the huge conference table surrounded almost entirely by middle-aged men. The camera focuses on a man, dressed in an immaculate business suit, who makes a passionate speech. "If we don't want television to founder totally in France, it's necessary to take a position," he says. "But not a stupid position of 'let's do this or that.' The state has to be conscious of television for culture, or better yet, for civilization." Then the video cuts to another scene with a man in a business suit, this time from the United States. "Well, I've made the comment that if the broadcasting of *Kojak* is going to be a major threat to your culture, then possibly you have a problem with your own culture" (*Distress Signals* 1990).

The New World Information and Communication Order debates of the 1970s and 1980s were complex, but it is worth keeping in mind the image of a conference room filled with professional men in business suits. Within the debates was a more grassroots-oriented tendency (including Raquel Salinas), but the movement was deeply contradictory. The debate started with the issue of unfair reporting of the Third World by news agencies based in London, Paris and New York. It soon broadened to include issues such as the allocation of the radio spectrum, lack of Third World access to telecommunication satellites, remote surveillance by the developed world of crops and minerals in the Third World, and many other complex issues. The debates took place within UNESCO, and an international committee was set up to write a report (MacBride 1980). The United States was hostile from the start and eventually withdrew completely from UNESCO (Preston, Herman and Schiller 1989). This effectively ended the movement, except for small groups of radical communication researchers who continued to meet from time to time (*Media, Culture & Society* 1990).

Debates within UNESCO included representatives of countries that had little interest in democratic media within their own country. The role of grassroots community media was only a small part of the discussion. The United States rejected any restrictions on its media industry in the world market. This included the computer industry and transnational flow. Clearly, the United States was already defending its dominant position in what would soon be called the "information age." Any restrictions on the free flow of information, on news and culture as commodities, were declared to be un-American and fiercely resisted. From a Latin American perspective, after a decade of military dictatorships and restrictions on freedom of expression, the issues looked quite different. The right to communicate was a fundamental democratic issue (Fox 1988).

Informativo *Tercer Mundo*

The tapes starts with music and then the announcement "*Tercer Mundo*: selections from international news for Latin America. Produced by Chasquihuasi Communications with the services of the news agency Inter Press Service. *Tercer Mundo*." There are eight items on tape #203. The first is from Brazil and concerns the decline in purchasing power due to the current economic crisis. Two professional broadcasters (one male, one female) read alternate sentences of the report, which is titled "In Brazil, the people continue to pay for the broken plates." There is a brief burst of Andean music and the next item is from Ecuador.

Runacunapac Yachana Huasi: The House of Knowledge

The campesinos of Simíatug in Ecuador have had their own radio for seven years: the radio Runacunapac Yachana Huasi. The radio functions in a small shack constructed by the campesinos in a *minga*. Its owner is eighteen indigenous communities. Radio Runacunapac Yachana Huasi is the only communication media that the indigenous people have in the Province of Bolívar to communicate among themselves and make their activities known.

When our radio doesn't transmit, it is as if one of our children has died, because it speaks about us and accompanies us in the morning and when we return home. This is how Manuel, member of one of the eighteen communities that own the radio, explains the importance of Radio Runacunapac Yachana Huasi.

The campesinos of Simíatug speak Quichua, and the programs teach how to improve the crops and take care of health, and tell legends and stories in their own language.

The transmission equipment of Radio Runacunapac Yachana Huasi was donated by another indigenous community eight years ago. The Shuar, indigenous to the region, upgraded their equipment and donated the old equipment to their brothers in Simíatug. In this way was born Radio Runacunapac Yachana Huasi.

When the Shuar donated their old equipment to us, nobody imagined that it would be one of the most important things here for a long time, said Samuel, who is in charge of the radio.

Every year there is a meeting of the eighteen communities to elect the eight people who work full-time at the radio. The work team of the radio has two coordinators in charge of planning and executing the programs.

On Saturdays and Sundays the radio transmits family festivities and visits the houses of listeners to record life—the life, art and hopes of the inhabitants of Simíatug. Everyone listens to the traditional healers. With every reason the campesinos of Simíatug call their radio Runacunapac Yachana Huasi, that is to say, the house of the people's knowledge.[3]

The item is expressed in clear and simple language. It explains that the equipment was obtained from another indigenous organization in Ecuador but that the building also involved collective work in a *minga*. It briefly explains how the station works. Two members of the community explain its value in their own words. Without getting into details of political debates and relations with outside institutions, the importance of the station is described as mainly cultural: it strengthens the people's knowledge.

Today We Are All Indispensable

It is best to evaluate the *Tercer Mundo* tapes in this way, with reference to concrete issues that have been explored earlier in the book. The reader will then be in a position to compare the two treatments. Here from tape #164 is a report on the Ecuadorian literacy campaign encountered in Chapter 2.

> The new government of Ecuador announced a massive campaign to teach reading and writing to all the population. The campaign carries the name of Monsignor Leonidas Proaño, who died on the 31st of August and whom the people called "the Bishop of the Indians."
>
> A hundred and seventy thousand volunteers will participate in the campaign, which has a cost of twenty million dollars. Among the literacy educators are brigades of secondary students, who can earn their school certificate by helping to teach literacy.
>
> In Ecuador fourteen per cent of the population, some 830,000 people, do not know how to read or write. The goal is to teach literacy to half a million illiterate persons in all of the country, including the most distant parts. Native cultures will also be respected, and literacy will be taught in Quichua and other native languages.
>
> The program has several stages. From now until January there will be seminars to train the literacy educators, prepare printed materials and open a campaign in the communication media. There will also be a campaign to collect twenty million dollars. The rest of the money will be paid by the United States and neighbouring countries to Ecuador.
>
> The work on the ground will start on the 15th of January when the volunteers begin to cover the country, teaching literacy. Every volunteer will bring printed materials about sanitation, nutrition and the defense of the environment to teach reading to everyone.
>
> The government of Ecuador counts on the participation of students, workers, the police, the Church, the Armed Forces, journalists and social organizations. The work will be a festival of unity and development of Ecuador, because political differences do not fit here, says the Minister of Education and Culture, Alfredo Veda.
>
> On the 15th of January, thousands of literacy workers will begin to cover the country united by one slogan: today we are all indispensable. We cannot wait longer—we will not wait longer.

This kind of item is central to the project of the *Tercer Mundo* tape service, which stresses concrete development campaigns and innovations relevant to rural Latin Americans. There is none of the skepticism about the motives of the government that we found among politicized indigenous organizations in Ecuador. It is clear from Chapter 2 that the presence of literacy workers from the army or the police in the indigenous communities would be completely inappropriate. Neither is there space within the format of *Tercer Mundo* for theoretical discussion of a campaign that uses standardized printed material—a far cry from the dialogical education advocated by Paulo Freire in similar circumstances in Brazil.[4]

A New Government for Old Problems

The following item concerns Bolivia and the 1989 political crisis that was described in Chapter 4. It allows for an evaluation of *Tercer Mundo* with respect to national politics within a specific Latin American country.

> Since Saturday the 5th of August, Bolivia has a new president, Jaime Paz Zamora. In this way ended ninety days of uncertainty about who will succeed Víctor Paz Estenssoro and govern Bolivia for four years.
>
> On the 7th of May, the Bolivian people went to the voting boxes to elect a president and parliament. But since none of the candidates got fifty-one per cent of the vote, it was up to the Congress to choose between the three candidates who got the most votes.
>
> In the elections in May, the largest vote went to Gonzalo Sánchez de Lozada of the official party, the Nationalist Revolutionary Movement (MNR). The second place was taken by Hugo Bánzer of the right-wing Nationalist Democratic Action (ADN), who was in charge of the dictatorship that governed Bolivia between 1971 and 1978. And in third place came Jaime Paz Zamora of the centrist Left Revolutionary Movement (MIR), who was vice-president of Bolivia between 1982 and 1985. In this way ended the uncertainty of Bolivians, which lasted ninety days.
>
> Jaime Paz Zamora included in his government various leaders of the National Democratic Action, the same party that had him imprisoned and exiled between 1971 and 1978. The new president of Bolivia, today moderate according to his own definition, in the past was a militant in the revolutionary left of Bolivia. Now he promises to maintain the free market economic system and combat the traffic of drugs during his term of office.
>
> In Bolivia it is known who will govern for the next four years. The new president of Bolivia is Jaime Paz Zamora. What we need to know is whether this government will resolve any of the problems that nobody else has solved for the past fifty years.[5]

This is a difficult matter to communicate in a short radio item. Listeners in Bolivia already know this information. Those in other countries may be interested, but it is not easy to explain the intricacies of Bolivian politics in a few minutes. Certainly the report is full of ironies and political turnabouts. What is most helpful is that the political parties are identified as "official" and "centrist" and "right-wing." However, this assumes these terms are significant for the listener. A previously leftist politician is now co-operating with the right wing. It is difficult to know what sense to make of this, and perhaps this explains the slightly cynical note on which the report ends.

Third World Realities

Overall, the *Tercer Mundo* tapes offer a worldview that is based less on a political ideology than on what Antonio Gramsci calls good sense. There is a coherent stance, but it is derived less from theoretical principles than from a progressive pro–Third World culture. *Tercer Mundo* can be quite outspoken about the need for substantial land reform, accompanied by all the necessary supports and decent prices for farmers' products. There is considerable interest in Third World countries' avoiding economic dependency—the need to pay for imports in dollars—by developing various forms of self-sufficiency. Along with attention to these macro issues, *Tercer Mundo* is equally interested in small development projects that make a difference to some people's lives. It pays considerable attention to issues of health and to relevant scientific discoveries. Its emphasis is more on the rural population and indigenous peoples—this is the explicit purpose of the service—than the city. There is also considerable attention to issues of women, children and persons with disabilities. In sum, it offers a Third World framework, but good sense rather than political theory.

The Chasquihuasi organization invited participating radio stations to send taped statements of how they use the cassette service. Among those who responded was Eusebio Sigcha of Radio Runacunapac Yachana Huasi in Simiatug. He points out that his small town is very remote and isolated from communication. The tapes from Santiago, Chile, woke them up to the reality of Latin America. They use the tapes in their information program. Their one commentary is that the voices go a little quickly. He points out that they speak Quichua and not everyone manages Spanish. But they give a summary in Quichua. The tapes are important: they reflect the reality in which we live. When we started to receive the cassettes from Chile, he says, we played the whole of one side of a tape, which is about twenty-five minutes. But that was too difficult for people to understand, and so now we use two or three items in our own news program as international news.[6]

Many other users of the cassette service repeat this pattern. The form allows for flexible uses in many different situations. Sometimes the items are translated into indigenous languages. News of Latin America is made relevant to local struggles. But the most frequent comment is that the news service allows listeners to situate their own lives and issues in a broader context. This is the central point made by Raymond Williams about the need always to understand local experiences in relation to other localities and to economic and political structures that press urgently on our lives but are only ever partly knowable.

1. The staff were reluctant to take on this responsibility, and the *Tercer Mundo* tape service was eventually moved to ALER in Quito.

2. The budget of the *Tercer Mundo* service in 1989 was US$30,000 per year, including rent and salaries.

3. *Informativa Tercer Mundo* #203 (sent on 21 July 1989). Translation by author.

4. The campaign materials emphasize dialogue and mutual respect and include discussion of Freire's ideas. However, the workbooks used by the students were based on ideas in the Universal Declaration of Human Rights (Campaña Nacional de Alfabetización 1988–89). This is an interesting choice and clearly raises many issues for discussion.

5. *Informativa Tercer Mundo* #207 (sent on 18 August 1989). Translation by author.

6. Chasquihuasi Comunicaciones, Empleo Tercer Mundo, 2nd part (1987–88).

Appendix A
Radio Latacunga[1]
Javier Herrán[2]

Radio Latacunga came on the air in December of 1981. Since then, the principal beneficiaries of this rich experiment in popular communication have been women's centers, youth groups, Christian assemblies and campesino centers, but fundamentally the Indigenous Movement of Cotopaxi in Ecuador.

Without knowledge or technical personnel, without experience in the field of social communication, the "Voice of the People on the March" emerged from groups involved in pastoral action in the Diocese of Latacunga.

Juanito and Luís, indigenous persons who for the first time pushed buttons and moved controls, were the first operators of the radio, along with Santiago, another indigenous person from Salcedo, who was in charge of the transmission equipment.

At first, everything seemed to indicate that between the technically sophisticated and those accustomed to the mattock there would not be the least understanding. The beginnings were a Calvary of blackouts and burnt-out diodes, of problems with the Harris equipment that didn't adapt to the altitude, of problems with the high-tension wires. These difficulties caused the indigenous people to say, "Radio Latacunga doesn't sleep, it seems as if someone has died in the high plateau." As soon as these difficulties had been surmounted, actual work was shaped, with much effort to promote fraternity and justice, to

---

1. This article was first published in *Materiales Para la Comunicación Popular* 9 (1987) published in Lima, Peru, by the Institute for Latin America (IPAL). Translation by Alan O'Connor.

2. A Catholic priest, Fr. Herran participated in the foundation and organization of Radio Latacunga in Ecuador.

capacitate the improvement of the living conditions of the campesinos and urban population, the improvement of production to satisfy the needs of the poorest.

Bilingual Programming

From the beginning, the necessity for two types of programming was clear: one in Spanish for the urban centers and Spanish-speaking campesinos, and another in Quichua for the indigenous communities. The first with the conventional shape of a commercial station, and the second, *Ricchari*, completely unedited and spontaneous in the voice of the protagonists themselves. This is the true Radio Latacunga. Here not only do the indigenous people express their life and culture in their own language and for their own people, but they themselves can form part of a "radio meeting" that is organized in the evening in the studio of Radio Latacunga at a frequency of 1080 AM. Two and one-half hours with a certain anarchic format, in which autochthonous music is mixed with greetings, communications, stories, news and denunciations.

Every afternoon numerous campesinos come to the radio station to say what's on their mind, to greet their families and to put in order judicial problems, forming an opinion in the community. Cassettes arrive with songs and music from bands. Campesino leaders turn up to convince neighbours of the benefits of whatever project is being organized and some to lash out at their adversaries.

The Quilotoa News Program

From the first meetings, the indigenous leaders decided on the necessity of a radio station to convoke the assemblies and to inform people about what is discussed in them. Their insistence counted for a lot in laying the foundations of Radio Latacunga. But this created a problem: the announcers who were the key figures at the radio station came to overshadow the leaders of the organizations. The power was in their hands, in the station, not in the organizations themselves. The radical solution arrived at was that the organizations of each zone would send to the station a program of news, communications, etc.

The first programs lasted for ten minutes but even so took more than an hour to record.

On Saturday, once the market is over, a group of community educators meet to put together the news of the parishes of Zumbahua, Guangaje and Chugchilán. The campesino leaders make contact with these and pass on the principal happenings of each community.

In this way the news of the campesinos' organizations is gathered. Then selections are made by the reporters of *El Noticiero Quilotoa* according to their significance and importance for each community.

Apprenticeship and Sociodrama

With an infrastructure that resembled the studios of a radio station, it was necessary to do things a little more "professionally." Before starting the work of recording, short explanations were given about some technical aspects.

Numerous exercises were done on the content of the news, its meaning and the different forms of telling it. The interview, for example, is a difficult technique to manage, even more so in a culture in which generally one wouldn't dream of getting directly to the "point." As well as this, "speaking to a microphone" is a very abstract idea for the campesinos. Operating the equipment was simpler than we had expected: imagination and good taste in music did the rest.

Even so, the problematic in which especially the indigenous communities of Zumbahua lived was not sufficiently reflected in *El Noticiero Quilotoa*. We then crossed over to the sociodrama, that is, the recreation of situations such as fiestas or wakes, and found in the dramatization the dynamic aspect of a social meeting. The preferred roles even today are the trader, the priest, the political boss, the patron—all of them linked to the suffering of the indigenous people.

The forestation program, the centers for campesino infants, the vaccination campaigns, the family orchard project, the training courses, they have found in the radio sociodrama an excellent way to reach potential participants.

The Extension of the Experience

The area covered by Radio Latacunga at present is more than the province [of Cotapaxi] thanks to the aid of government and private institutions.

Today there are seven campesino organizations that have a recording cabin in their social network: in Zumbahua, the Union of Community Councils of Zumbahua and Indigenous Schools of Quilotoa, with thirty-five communities; in Salcedo, the Campesino House of Salcedo, with twenty communities; in Cusubamba, the Greater Council of Cusubamba, with seventeen communities; in Toacazo, the Union of Campesino Organizations of the North of Cotopaxi, with fifteen communities; in Pujilí, the Campesino House of Pujilí, with fifty communities; in Saquisilí, the Campesino House of Saquisilí, with fifteen communities; in Chugchilán, the Union of Community Councils of Chugchilán, with eight communities.

The challenge is here. Popular participation gives rise to more participation and other sectors reclaim their space here: the Federation of Neighborhoods of Latacunga and the Federation of Secondary Students of Cotopaxi.

As the experience has grown in numbers, it has become necessary to attend more to the end product.

With the collaboration of institutions such as CIESPAL, Radio Netherlands and the Ministry of Education and Culture, the campesino organizations have been able to train many of their members who work as volunteers in the recording cabin of their organization. This training, as much as on-site as in courses, seeks to balance knowledge and various techniques that permit the popular reporters to be better communicators without losing the naturalness and freshness of their expression. Frequently the variety of contents of a program has caused the message to be diluted. In that way the necessity for training, more than the transmission of radio techniques, is a continual dialogue of why and for what in the process of popular communication.

The Campesino Radio Magazine

News, communications, sociodrama, the illustrative dialogue or chat and commentary in the form of editorials, together with music by local groups and bands, riddles, greetings, interviews, stories and agricultural advice, express the life of the community.

The work that makes direct actors of the campesinos themselves determines the search for new radio techniques that permit live and spontaneous expression and at the same time establish continuity in the development of the theme.

It is the people who communicate their wisdom in surviving adverse ecological and social conditions that development projects do not have the capacity to alter.

In this way, then, the contents of the "campesino magazine" are in turn religion, health, cooking, agriculture, animal sanitation, forestry, artisan crafts, erosion, autochthonous music, native culture, Spanish language, law, etc.

The Difficult Road

It has not been possible to complement social mobilization and the growth of what is needed with a parallel process of self-finance. Campesino organizations do not have the capacity to finance the operating costs that Radio Latacunga accumulates.

As well as this, the economic situation of the radio station does not allow an increase in the number of services and persons needed by the social development movement, which in turn created the recording cabins of the campesino organizations.

The time has come for institutions that are concerned with participatory communication to study this experience and establish possibilities to strengthen it and repeat it in other radio stations with the vocation of service to marginalized peoples.

Appendix B
Aymaras and Christians[1]

We Aymaras have lived in this land since time immemorial; there are no documents about this. We have our history and religion in the rocks, in our bones, in weavings and in our mind. Since that time we have maintained the same form of life and kept the same religion. The potato is ours, quinua, maize and the llamas, and we even go to the high places to ask the Father God to guide and protect us. We have a well-established religion, well directed and in harmony with the life of the community. Nowadays, however, things are rapidly changing: the school, army barracks and university direct development to destroy what is natural. These all make an attempt against our culture, because for the Aymaras there is no development without respect for the land. For this reason we must direct ourselves well and turn to worship our Father the Sun, who germinates life.

We are a people who have created our food. Our culture consists of knowing and making good use of nature without disfiguring its face. For example, coca for us is the same as the Bible for you. It is with us at all moments: at fiestas and wakes, when we are well and when we are sick, when we are about to prepare the land for seed and when we relax. Mother Coca is always with us; with her we live.

This is the greatness of the Aymaras: we consider all things mutually related, that everything has to be seen as a whole and neither anything nor anybody is isolated. This greatness must not disappear; it must grow so that we

---

1. Extracts from commentaries by four important Aymara leaders in a workshop on Aymara religion and Christianity sponsored by the Center for Popular Theology in La Paz, Bolivia. Taken from the magazine *Fe y Pueblo* and reproduced in *Materiales Para la Comunicación Popular* 9 (1987) published in Lima, Peru by the Institute for Latin America (IPAL). Translation by Alan O'Connor.

may resist the misery that has brought us a vision of the world that sees things as isolated and without responsibility to their surroundings.

The Aymara religion will not be lost. Some of our brothers know the Bible, but also carry out service to our deities. Our religion is one of mutual caring among the people and care for that which surrounds us: the fields, the animals, the high altitudes. The offering to the Pachamama (Mother Earth) must be made with much affection, from the bottom of our heart.

The *yatiri* is the curer, the magician, the fortune-teller. He is a person selected by God, who after being touched by the ray goes to the high altitudes to learn from the eldest, from those most experienced.

At the beginning of the colonial period, the Aymaras were obliged to convert to Christianity. In order to preserve our religion in the past, our ancestors had to mix it with the Christian religion.

In the most remote places, the Christian religion is not accepted. Included here are the places the missionaries managed to penetrate but did not succeed in all of their aims, and now our brothers accept the Catholic teachings in only a superficial manner. We have been put out of joint by the death of the Inca; to accept the Christian religion is a way to protect ourselves and continue being ourselves. In this way, for example, after being born, a child is baptized not because of what the missionary says but so as not to provoke a hailstorm or a frost that would destroy our crops.

It could be said that even though they are considered to be Christian, in the bottom of their heart they continue to believe in our religion; it can not be separated from our culture.

For us the Christian religion can be of use. It is good to read the Bible. We consider it to be something theoretical and our own religion as practical. The two tend to complement each other. The first religious who arrived here did not spread the word of God. They were unusual people among whom there were some women. Then everything changed. Priests began to come and began to instruct, teach literacy and undertake social promotion projects. We believe that this aspect is positive but that all of the religious aspect has been lost. It seems that the new religion is the market; it is desired that through it we will be westernized.

Aymara Justice

In 1984, the Aymaras had a very important meeting. We discussed our economic problems and in this respect decided that anyone who had a hundred llamas must divide them, so that the number is reduced, to the others who only have forty or fifty. This we call *kuskachaña*, which means to level. This is a very important point in the Aymara vision of the economy and is the opposite of your vision, which favours accumulation to the detriment of others.

We ask ourselves why, in the countries that are called developed, factories have been installed and autos constructed that contaminate the plants, the woodlands and every living being. This is suicide and not progress.

A Divided Church

We have the impression that Christianity is in crisis. There are great contradictions between the different confessions: the Protestants say one thing and the Catholics another. They speak of love but they hate; speak of unity but they are divided; speak of community but they are individualists: in most cases they speak of community of faith only in order to share the least possible.

If we have understood it, in Christianity there should not be rich and poor, but we know that the rich and powerful of this world are Christians. There is a falsehood at the bottom of Christianity that is hidden by words. Aymaras cannot accept religions with these kinds of discrepancies.

We are aware that now we cannot banish the Christian religion, which has been established here for a long time. The Christians are moreover the owners of the communications media and all of the instruments of power. The one thing we ask is that we be respected, but for this the Christians must become poor and give up their power. Because we also know that Christianity has been in the arms of the military and economic forces. These infrastructures are needed for evangelical work, and this always carries with it the subjugation of others, although it is pretended that this is not the case.

You should evangelize in the cities to the Christians themselves, above all to the government members who are Catholics. Leave us in peace; we already have our religion and for us it is a good religion. You concern yourselves with being good Christians, and we will concern ourselves with being good Aymaras.

Bibliography

Books

Adorno, Theodor. 2000. *The Psychological Technique of Martin Luther Thomas' Radio Addresses*. Stanford: Stanford University Press.

ALER. 1983. *Hacia Una Comunicación Participativa: Entrevista a Mario Kaplun*. Quito: Asociación Latinoamericana de Educación Radiofónica.

Barrios de Chungara, Domitila. 1978. *Let Me Speak! Testimony of Domitila, a Woman of the Bolivian Mines*. With Moema Viezzer. Trans. Victoria Ortiz. New York and London: Monthly Review Press.

Beltran, Luis Ramiro. 1980. "A Farewell to Aristotle: Horizontal Communication." *Communication* 5:5–41.

Brecht, Bertolt. 1983. "Radio as a Means of Communication: A Talk on the Function of Radio." In *Communication and Class Struggle, 2. Liberation, Socialism*, ed. Armand Mattelart and Seth Siegelaub. New York: International General.

Campaña Nacional de Alfabetización. 1988–89. *Documento de Trabajo*, nos. 1–21. Quito.

CIESPAL. 1988. *Directorio de Institutiones con Proyectos de Comunicacion y Educación Popular en Ecuador*. Quito: CIESPAL.

CINCO. 1987. *Comunicación Dominante y Comunicación Alternativa en Bolivia*. La Paz: CINCO.

Colloredo-Mansfield, Rudi. 1999. *The Native Leisure Class: Consumption and Cultural Creativity in the Andes*. Chicago: University of Chicago.

Corkill, David and David Cubitt. 1988. *Ecuador: Fragile Democracy*. London: Latin American Bureau.

Cornejo Portugal, Inés. 1996. La Radio Cultural Indigenista: Algunas Respuestas." In *La Perspectiva de Etnias y Naciones: Los Pueblos Indios de América Latina*, ed. Salomón Nahmad. Quito: Ediciones Abya-Yala.

Dandler, Jorge. 1976. "Peasant Sindicatos and the Process of Cooptation in Bolivian Politics." In *Popular Participation in Social Change: Cooperatives, Collectives, and Nationalized Industry*, ed. June Nash, Jorge Dandler, Nicholas S. Hopkins. The Hague: Mouton.

———. 1983. *Sindicalismo Campesino en Bolivia: Cambios Estructurales en Ucureña, 1935–1952*. La Paz: CERES.

de Vela, Gloria. 1988. "Diagnóstico Comunicacional de las Organizaciones Campesinos de Cotopaxi." In *Tres Experiencias de Diagnóstico de Comunicación*, ed. Daniel Prieto Castillo. Quito: CIESPAL.

Dunkerley, James. 1984. *Rebellion in the Veins: Political Struggle in Bolivia, 1952–1982*. London: Verso.

Dunkerley, James and Rolando Morales. 1986. "The Crisis in Bolivia." *New Left Review* 155:86–106.

Fanon, Frantz. 1965. "This is the Voice of Algeria." In *A Dying Colonialism*. Trans. Haakon Chevalier. New York: Grove Press.

Field, Les. 1991. "Ecuador's Pan-Indian Uprising." *NACLA Report on the Americas* 25 (3): 39–44.

Forgacs, David, ed. 1988. *An Antonio Gramsci Reader: Selected Writings, 1916–1935*. New York: Schocken Books.

Fox, Elizabeth, ed. 1988. *Media and Politics in Latin America: The Struggle for Democracy*. London: SAGE Publications.

Freire, Paulo. 1973. "Extension or Communication." Trans. Louise Bigwood and Margaret Marshall. In *Education for a Critical Consciousness*. New York: Continuum.

Galeano, Eduardo. 1998. *Soccer in Sun and Shade*. London and New York: Verso.

Garcia, Dennis. 1985. Una Voz en la Boca del Lobo. Unpublished paper, CEDECO, Quito.

Gill, Lesley. 2000. *Teetering on the Rim: Global Restructuring, Daily Life, and the Armed Retreat of the Bolivian State*. New York: Columbia University Press.

Goffin, A. M. 1994. *The Rise of Protestant Evangelism in Ecuador, 1895–1990*. Florida: University Press of Florida.

Goffman, Erving. 1981. "Radio Talk." In *Forms of Talk*. Philadelphia: University of Pennsylvania Press.

Gramsci, Antonio. 1971. *Selections from the Prison Notebooks*. Trans. Quintin Hoare and Geoffrey Nowell Smith. New York: International Publishers.

———. 1981. "Antonio Gramsci." In *Culture, Ideology and Social Process: A Reader*, ed. Tony Bennett and others. London: Batsford.

Gumucio Dagron, Alfonso. 1982. "El Papel Politico de las Radios Mineras." *Comunicación y Cultura* 8:89–99. Trans. in O'Connor 2004.

———. 2000. *Making Waves: Stories of Participatory Communication for Social Change*. New York: Rockefeller Foundation.

Gumucio Dagron, Alfonso and Lupe Cajías, eds. 1989. *Las Radios Mineras de Bolivia*. La Paz: CIMCA.

Gwyn, Richard J. 1983. "Rural Radio in Bolivia: A Case Study." *Journal of Communication* 33:79–87.

Hale, Charles. 1994. "Between Che Guevara and Pachamama." *Critique of Anthropology* 14:1.

Hein, Kurt. 1984. "Popular participation in rural radio: Radio Bahá'í, Otavalo, Ecuador." *Studies in Latin American Popular Culture* 3:96–104.

Hein, Kurt John. 1988. *Radio Bahá'í, Ecuador: A Bahá'í Development Project*. Oxford: George Ronald.

Herrán, Javier. 1987. "Radio Latacunga." *Materiales Para la Comunicación Popular* 9.

Hornik, Robert C. 1988. *Development Communication: Information, Agriculture and Nutrition in the Developing World*. New York and London: Longman.

Hoxeng, James. 1977. Programming by the People: An Ecuadorian Radio Experiment." *Educational Broadcasting International* (March): 30–34.

Huesca, Robert. 1995. "A Procedural View of Participatory Communication: Lessons from Bolivian Tin Miners' Radio." *Media, Culture & Society* 17:101–119.

———. 1997. "Low-Powered Television in Rural Bolivia: New Directions for Democratic Media Practice." *Studies in Latin American Popular Culture* 16:69–90.

Hvalkov, Søren and Peter Aaby, eds. 1981. *Is God an American? An Anthropological Perspective on the Missionary Work of the Summer Institute of Linguistics*. Copenhagen and London: IWGIA and Survival International.

Ibarra, Alicia. 1987. *Los Indigenas y el Estado en el Ecuador: La Práctica Neoindigenista*. Quito: Ediciones Abya-Yala.

Instituto Indigenista Interamericano. 1986. *Seminario-Taller Sobre Radiodifusión en Regiones Indígenas de America Latina: Informe Final*. Quito: UNESCO, CIESPAL, ICI, UNDA-AL, ALER.

———. 1987. "Hacer la Radio a Diario: Memoria del II Seminario-Taller sobre Radiodifusión en Regiones Indígenas de América Latina." *Anuario Indigenista* 47:11–71.

———. 1988. "Para Mantenernos en Contacto . . . Secretaria Permanente: Relación del III Seminario-Taller sobre Radiodifusion en Regiones Indígenas de América Latina." *Anuario Indigenista* 48:27–32.

Iza, Dioselina. 1986. "La Radio y la Organización Campesina." In Instituto Indigenista Interamericano 1986.

Katz, Elihu and Paul Lazarsfeld. 1955. *Personal Influence: The Part Played by People in the Flow of Mass Communication*. Glencoe: Free Press of Glencoe.

Korovkin, Tanya. 1997. "Taming Capitalism: The Evolution of the Indigenous Peasant Economy in Northern Ecuador." *Latin American Research Review* 32 (3): 89–110.

Kúncar Camacho, Gridvia. 1989. *Comunicación Alternativa y Sindicalismo en Bolivia: La Experiencia de las Radio Mineras (1950 a 1980)*. La Paz: CEBIAE.

Lagos, Maria L. 1994. *Autonomy and Power: The Dynamics of Class and Culture in Rural Bolivia*. Philadelphia: University of Pennsylvania Press.

Latin American Bureau. 1987. *The Great Tin Crash: Bolivia and the World Tin Market*. London: Latin American Bureau.

Lazarsfeld, Paul F. 1946. *The People Look at Radio*. Chapel Hill: University of North Carolina Press.

Lerner, Daniel. 1958. *The Passing of Traditional Society: Modernizing the Middle East*. New York: Free Press.

Lewis, Oscar. 1961. *The Children of Sánchez: Autobiography of a Mexican Family*. New York: Vintage Books.

Llorens, José A. 1991. "Andean Voices on Lima Airwaves: Highland Migrants and Radio Broadcasting in Peru." *Studies in Latin American Popular Culture* 10:177–90.

López Vigil, José Ignacio. 1985. *Radio Pío XII: Una Mina de Coraje*. Quito: ALER/Pío XII.

Löwy, Michael. 1996. *The War of Gods: Religion and Politics in Latin America*. London and New York: Verso.

Lozada, Fernando and Gridvia Kúncar. 1983. "Las Emisoras Mineras en Bolivia: Una Histórica Experiencia de Comunicación Autogestionaria." In *Comunicación Alternativa y Búsquedas Democráticas*, ed. Fernando Reyes Matta. Santiago, Chile: ILET. Trans. in O'Connor 2004.

MacBride, Sean. 1980. *Many Voices, One World: Report by the International Commission for the Study of Communication Problems*. London: Kogan Page.

Marcos, subcomandante. 1995. *Shadows of Tender Fury*. Trans. Frank Baracke and others. New York: Monthly Review Press.

Marcus, George E. 1998. *Ethnography through Thick and Thin*. Princeton: Princeton University Press.

Marcus, George E. and Michael M. J. Fischer. 1986. "Taking Account of World Historical Political Economy: Knowable Communities in Larger Systems." In *Anthropology as Cultural Critique: An Experimental Moment in the Human Sciences*. Chicago and London: University of Chicago Press.

Materiales para la Comunicación Popular, no. 9 (1987). Published by IPAL in Lima, Peru.

Michaels, Eric. 1994. *Bad Aboriginal Art: Tradition, Media, and Technological Horizons*. Minneapolis: University of Minnesota Press.

Miles, Ann. 1998. "Radio and Commodification of Natural Medicine in Ecuador." *Social Science and Medicine* 47 (12): 2127–2137.

Mintz, Sidney. 1985. *Sweetness and Power: The Place of Sugar in Modern History*. New York: Penguin Books.

Muratorio, Blanca. 1981. "Protestantism, Ethnicity, and Class in Chimborazo." In *Cultural Transformations and Ethnicity in Modern Ecuador*, ed. Norman E. Whitten Jr. Urbana: University of Illinois.

Musto E. 1971. *Communication Media for Rural Development: The Colombian Model "Radio Sutatenza."* Bogotá: Acción Cultural Popular.

Nash, June. 1979a. "Ethnology in a Revolutionary Setting." In *The Politics of Anthropology: From Colonialism and Sexism Towards a View from Below*, ed. Gerrit Huizer and Bruce Mannheim. The Hague and Paris: Mouton Publishers.

———. 1979b. *We Eat the Mines and the Mines Eat Us: Dependency and Exploitation in Bolivian Tin Mines*. New York: Columbia University Press.

O'Connor, Alan. 1989a. "Bolivian Miners' Radio: Conflicting Solutions to Crisis." *Inter/Radio: The Newsletter of the World Association of Community Radio Broadcasters* 2 (2): 1, 10.

———. 1989b. "Bolivian Workers Fight to Keep Miners' Radios on the Air." *Labor Notes* 118 (Jan): 1, 14.

———. 1989c. "Economic Threat to Bolivian Miners' Radio Network," *The Democratic Communiqué* (Winter): 8.

———. 1989d. *Raymond Williams: Writing, Culture, Politics*. Oxford: Basil Blackwell.

———. 1990a. "Between Culture and Organization: The Radio Stations of Cotopaxi, Ecuador." *Gazette: The International Journal for Mass Communication Studies* 46 (August): 81–91.

———. 1990b. "Materials for a Comparative Analysis of Alternative Media in Ecuador and Bolivia." *Communicatio Socialis Yearbook* (India): 53–83.

———. 1990c. "Miners' Radio Stations in Bolivia: A Culture of Resistance." *Journal of Communication* 40 (1): 102–110.

———. 1990d. "Radio is Fundamental to Democracy," *Media Development* 27 (4): 3–4.

———. 1991a. "The Alternative Press in Bolivia and Ecuador: The Examples of *Aqui* and *Punto de Vista.*" *Howard Journal of Communication* 2 (4): 349–356.

———. 1991b. "The Emergence of Cultural Studies in Latin America." *Critical Studies in Mass Communication* 8:60–73.

———. 1993. "People's Radio in Latin America." In *Progress in Communication Sciences* 11:207–28.

———, ed. and trans. 2004. *Community Radio in Bolivia: The Miners' Radio Stations.* Pref. by John Downing. Lewiston, NY: Edwin Mellen Press.

O'Sullivan-Ryan, Jeremiah and Mario Kaplun. 1978. *Communication Methods to Promote Grass-Roots Participation: A Summary of Research Findings from Latin America.* Paris: UNESCO. ERIC Reports document no. 201017.

Pachano, Simon. 1986. *Pueblos de la Sierra.* Quito: Programa de Investigaciones Sociales sobre Población en América Latina and Instituto de Estudios Ecuatorianos.

Preston, William Jr., Edward S. Herman and Herbert I. Schiller. 1989. *Hope and Folly: The United States and Unesco, 1945–1985.* Minneapolis: University of Minnesota Press.

Price, Richard and Sally Price. 1992. *Equatoria.* New York and London: Routledge.

Price, Sally and Jean Jamin. 1988. "A Conversation with Michel Leiris." *Current Anthropology* 21 (1): 157–74.

Radio Fides. 1989. Radio Fides 50 Años 1939–1989. *Presencia.* January 29.

Reyes Matta, Fernando. 1986. "Alternative Communication: Solidarity and Development in the Face of Transnational Expansion." In *Communication and Latin American Society: Trends in Critical Research, 1960–1985,* ed. Rita Atwood and Emile G. McAnany. Madison: University of Wisconsin Press.

Rivadeneira Prada, Raúl. 1982. *Resistencia y Coexistencia: Cultura Boliviana y Cultura Transnacional.* La Paz: Editorial Gisbert.

Ruiz, Carmen. 1994. "Losing Fear: Video and Radio Productions of Native Aymara Women in Bolivia." In *Women in Grassroots Communications: Furthering Social Change,* ed. Rilar Riano, 161–78. Thousand Oaks: SAGE.

Salinas, Raquel. 1984. *Agencias Transnacionales de Información y el Tercer Mundo.* Quito: CIESPAL.

Samarajiwa, Rohan. 1987. The Murky Beginnings of the Communication and Development Field: Voice of America and "The Passing of Traditional Society." In *Rethinking Development Communication,* ed. Neville Jayaweera and Sarath Amunugama. Singapore: Asian Mass Media Communication Research Centre.

Scannell, Paddy, ed. 1991. *Broadcast Talk.* London: SAGE Publications.

Schmelkes de Sotelo, Sylvia. 1973. *The Radio Schools of the Tarahumara, Mexico: An Evaluation.* Washington: Academy for Educational Development.

Schmucler, Héctor and Orlando Encinas. 1982. "Las Radios Mineras de Bolivia (Entrevista con Jorge Mancilla Romero)." *Comunicación y Cultura* 8:69–86. Trans. in O'Connor 2004.

Schramm, W. 1977. *Big Media, Little Media: Tools and Technologies for Instruction.* Beverley Hills, London: Sage.

Sparks, Colin and Colleen Roach, eds. 1990. *Media, Culture & Society.* Farewell to NWICO? Special issue. 12 (July).

Taussig, Michael. 1980. *The Devil and Commodity Fetishism in South America.* Chapel Hill: University of North Carolina Press.

———. 1987. *Shamanism, Colonialism and the Wild Man: A Study in Terror and Healing.* Chicago and London: University of Chicago Press.

Taylor, Anne-Christine. 1981. "God-Wealth: The Achuar and the Missions." In *Cultural Transformations and Ethnicity in Modern Ecuador,* ed. Norman E. Whitten Jr. Urbana: University of Illinois.

Torres, Camilio. 1971. "The Radio Schools of Sutatenza." In *Revolutionary Priest: The Complete Writings and Messages of Camilo Torres,* ed. John Gerassi. New York: Random House.

Vandenbulcke, Humberto. 1986. "Radio Runacunapac Yachana Huasi, Simiatug, Ecuador." In Instituto Indigenista Interamericano 1986.

Vargas, Lucila. 1995. *Social Uses and Radio Practices: The Uses of Participatory Radio by Ethnic Minorities in Mexico.* Boulder: Westview Press.

Weinstock, Steven. 1973. The adaption of Otavalo Indians to urban and industrial life in Quito, Ecuador. Ph.D. dissertation, Cornell University

Whitten, Norman E. Jr. 1976. *Ecuadorian Ethnocide and Indigenous Ethnogenesis: Amazonian Resurgence Amidst Andean Colonialism.* Copenhagen: International Working Group for Indigenous Affairs.

Whitten, Norman Jr. 1985. *Sicuanga Runa: The Other Side of Development in Amazonian Ecuador.* Urbana and Chicago: University of Illinois Press.

Whitten, Norman, Dorothea Scott Whitten and Alfonso Chango. 1997. "Return of the Yumbo: The Indigenous Caminata from Amazonia to Andean Quito." *American Ethnologist* 24 (2): 355–91.

Whittington, Bruce. 1985. Some Latin American community radio experiments: A report to CUSO. Unpublished report. Ottawa.

Williams, Raymond. 1974. *Television: Technology and Cultural Form.* Glasgow: Fontana.

———. 1975. *The Country and the City.* St Albans: Paladin.

———. 1976. *Communications.* Harmondsworth: Penguin.

———. 1977. *Marxism and Literature.* Oxford: Oxford University Press. In Spanish, *Marxismo y Literatura,* trans. Pablo di Masso. Barcelona: Ediciones Península.

Wolf, Eric R. 1982. *Europe and the People without History.* Berkeley: University of California Press.

Wright, Ronald. 1986. *Cut Stones and Crossroads: A Journey in Peru.* New York: Penguin.

Zamosc, L. 1994. "Agrarian protest and the Indian movement in the Ecuadorian highlands." *Latin American Research Review* 21 (3): 37–69.

Videos

Boca de Lobo. c. 1985. R. Khalifé and Susana Andrade. Quito: Casa de la Cultura Ecuatoriana.

Distress Signals: An Investigation of Global Television. 1990. Dir. John Walker. Ottawa: National Film Board of Canada.

Luis R. Beltran Tunes Into Bolivian Miner's Radio. 1991. New York: Paper Tiger Television.

Marketing Ideas, Selling Nutrition. 1980. Washington: USAID.

The Voice of the Mines. 1984. Dir. Eduardo Barrios and Alfonso Gumucio Dagron. Paris: UNESCO.

Index

About the Author

Alan O'Connor received his first degree in sociology at Trinity College, Dublin, and was awarded his Ph.D. by York University in Toronto for his research on Raymond Williams. Dr. O'Connor has taught at Ohio State University and is now associate professor in the Cultural Studies Program at Trent University in Canada. A founder of *Borderlines* magazine and the Toronto bookstore Who's Emma, he has published extensively on community media in Latin America. His most recent books are *Community Radio in Bolivia: The Miners' Radio Stations* (Mellen, 2004) and *Raymond Williams* (Rowman and Littlefield, 2005). He is currently researching a book on punk subculture in Mexico, Spain and Canada.